The Crossroads of Social and Climate Justice

An Exploration of Issues & Solutions for Planet and People

The Crossroads of Social and Climate Justice

An Exploration of Issues & Solutions for Planet and People

LaVerne Hillis McLeod

Purple Feather Press

Cover painting © Seema Christie
Editing: Dianne Miller, George Somero, Maria Byford,
Ken McLeod & Alison Mitchell
Created with Vogel Graphics
(Formatting, E-book and Printing Conversions)
First published in December 2022

Published by Purple Feather Press
Post Office Box 566
Big Sur, California 93920 USA

Library of Congress Cataloging-in-Publication Data

The Crossroads of Social and Climate Justice:
An Exploration of Issues & Solutions for People and Planet

Includes bibliographical references.

ISBN: 978-0-9976531-5-1 (Paperback)

ISBN: 978-0-9976531-6-8 (E-book)

First Edition

Printed by Ingram-Spark USA

DEDICATED TO:

To Lynda Sayre who inspired me to focus on climate issues that affects people and world conditions.

To the late Denyse Frischmuth for her unbending commitment with Communities for a Sustainable Monterey County (CSMC) to promote efforts to heal Planet Earth

To Helen Rucker, a pioneer for racial justice. She has always supported me from my inception as an activist for social change to current times.

To my grandchildren, Toni Breighann McLeod, Kenneth William McLeod II, Penelope Aurora McLeod and any grandchildren born after this publication. They are the ones who will inherit people and planet issues. I wish them to carry the torch for positive change.

To my husband, Kenneth McLeod, who tolerated with my very early morning awakening to write. I am grateful for his support.

To my son, Tobias McLeod, who boosted my spirits each time I talked about my writing journey.

To everyone of all ages, who want to explore possible career, project or resource information to help reduce greenhouse gases and establish equity and reparations on planet Earth.

CONTENTS

SECTION III
EARTH

Chapters   Page

The Elemental Mirror of Earth

SECTION IV
FIRE

The Elemental Mirror of Fire

SECTION V

You Might Consider:

i. **Acknowledgements**

As songwriter Steph Drouin wrote these lyrics, both she and Aimee Ringle so beautifully sang:

> I'm not a lone wolf and I never was
> Anything I achieve I achieve it because
> I am standing on the shoulders
> Of an infinite many seen and unseen ...

Lynda Sayre who inspired me to join B-SAGE (Big Sur Advocates for a Green Environment) otherwise the idea to write this book might not have emerged.

Denyse Frischmuth (posthumously) Who inspired me to continue with climate justice work as I was impressed with her relentless passion and leadership with CSMC (Community for a Sustainable Monterey County).

Big Sur Fashion Show that allowed me free expression of theatrical modeling designs that echoed solutions for the various continent consequences caused by worldwide heat rise.

Dr. George Somero for his professional scientific marine biology knowledge and his guidance as an interviewee.

Seema Christie, for her symbolic painting that was inspiration for the cover of this book.

Janna Ratzlaff, Justin Wright and **James Galvin** for sharing their aquaponics business insights through personal interviews.

Dennis Burns, Fire Behavior Analyst, who kindly allowed me to quiz and interview him about his knowledge in studying fires.

Steve Sanders, Geologist, who simply explained how he took part in a tree planting organization.

Jessica Koning, a Big Sur, CA biologist, for enthusiastically providing some research on Biomimicry to assist with flooding solutions.

Walden Kiker, environmental researcher and student at Monterey Institute of International Studies (MIIS) that helped in various aspects of this book.

Ed Van Weijen, for his Soberanes Fire photography.

Ben Eichorn, for his contribution with intersecting the climate crisis with human patterns of consumption and waste.

Kari Bernardi, for her contribution with food choices harmonizing with climate and social justice.

Susan Olesek, for her willingness to allow me to share a story about the human potential movement for prisoners called The Enneagram Prison Project (EPP). A chapter highlights a person who benefitted from this project, step by step.

And lastly, I give credit to the divine spiritual energy that kept me connected to my purpose for writing this book. That purpose is to help inspire readers to be called to action through suggested solutions given for various social and climate justice issues.

ii. **Forward**

As the severity of the crisis of anthropogenic (man-made) global change increases, the number of issues being examined by different types of scholars is growing. In the earlier years of study of global change, scientists focused primarily on what might be termed the "physics" of the phenomenon, namely the causes of the rise in temperature that has been recorded since the onset of the Industrial Revolution in the early to middle part of the 19th century. The emission of greenhouse gases, notably carbon dioxide (CO_2) was shown to be the proverbial "smoking gun" behind the warming trend. Physicists, geologists and oceanographers then added a focus on loss of ice and the rise in sea level, processes that portend enormous consequences for the biosphere. As the severity of these physical threats was appreciated, scientists began to increasingly examine the biological consequences of warming. How would warming affect the structures of ecosystems and the distribution patterns of species, plant and animal.

In my own work, emphasis was on the underlying physiological effects of temperature: How do the different organs, cells and macromolecules of organisms respond to rising temperature—and what types of adaptive changes might be possible to offset some of these effects.

Even though these broad studies of the physics, chemistry and biology of global change did include consideration of impacts on human populations, the discussions of challenges to human societies were generally very generic and involved impacts on agriculture, problems for cities arising from sea level rise, and challenges in maintaining a stable body temperature in a hotter and more humid climate. To a large degree, the humanity was discussed as a whole; there was inadequate attention given to the huge diversity of impacts of global change on people of different nations, racial groups, cultures, and economic status. Thus, to adequately comprehend the horrific problems arising from global change, a more fine-scale analysis of effects on humans was needed, one that included a central focus on critical aspects of social justice. We are not all equally threatened by the on-going changes we have inflicted on our environment. Rather, some populations, notably certain racial and economic groups are destined to bear a disproportionate level of stress from global change.

Ms. McLeod's book takes on the challenge of integrating the effects of global change on the environment at large—increasing temperatures, sea level rise, increased fire danger, reduction in agricultural production and other effects—with social justice issues that spell out the differential levels of these stressors on different groups of people. She has truly done her homework in creating this synthetic analysis! As her text and list of references indicates, she has mastered a wide range of literature dealing with the fundamental physics, oceanography, and ecology of global change, and paired this expertise with an equivalent deep study of how these environmental changes will play out among different racial and economic groups. This is a synthesis that has been begging to be done, and Ms. McLeod has more than risen to this task. She not only presents a solid factual foundation for all of the analyses she performs, but she also provides the reader with information on how to access information across the full spectrum of topics she studies. Importantly, she includes many sources of information about programs and organizations that are working to mitigate the effects of global change. Thus, she tells us about the seriousness of the environmental and social problems we face and then, points out positive actions that we can take to address them. This is a very important service to her readers. The facts about global change cannot help but make us feel demoralized. By pairing these startling and sad facts with information on how to help resolve the problems she discusses, she allows her readers to emerge with a sense of optimism that each of us has many pathways open for doing deeds that can reduce the severity of the problems she presents. We owe her a great deal of thanks for helping us to at once understand the problems we face and to appreciate the many positive acts we can—and definitely should—perform to undo the damage that our species has done to the biosphere of which we are a part.

by: Dr. George Somero,
The David and Lucille Packard Professor of
Marine Science Emeritus currently working at
Stanford University's Hopkins Marine Station

iii. Introduction by the Author

My worldly concerns have revolved around climate and people. This interest has been intensified as an educator, environmentalist and a social justice advocate. Currently, I am co-coordinator of the environmental group, Big Sur Advocates for a Green Environment (B-SAGE). I am also creator and facilitator of Bridge Building to Equity's Workshop and Webinar Series. ©2013. My focus with both of these endeavors is the crossroads or intersection from which I wrote this book. It's about the place where we can sit with a mirror that reflects our current societal realities. It is also a place to explore the issues and suggestions for dissolving them.

I was guided by a question and a theatrical/modeling performance to inspire me to write this book. The question is an empowering and motivating one by Dr. Martin Luther King, Jr:
"Life's most persistent and urgent question is, what are you doing for others?"
With this question, I became profoundly concerned about my brown-skinned grandchildren as heirs of this planet. Being deeply affected by the social injustice of people of color in the United States and the world, writing *The Crossroads of Social and Climate Justice* answered the question.

Secondly, being motivated after a sizzling hot-weather performance presented through theatrical/modeling called "Signs of the Time," I connected the dots of climate consequences affecting people. I designed clothing for this performance in the Big Sur Fashion Show for seven models representing each continents' heat wave and am grateful for this prompting.

Even though, my writing mostly represents United States examples, it applies to other countries. The levels of severity can vary. I can therefore conclude that this book is written for the purpose of inspiring action to help decrease the amount of greenhouse gases in planet earth's atmosphere. Combined with this action is finding ways or solutions to help dissolve racial and systemic injustices.

Now is the time to continue moving forward into change- a movement to heal both our earth and its suffering inhabitants by creating a bridge to connect and heal together.

The Crossroads of Social and Climate Justice, addresses some issues that throw light on a few "elephants in the room." This delivery is an opportunity to understand and hopefully activate action to change something rather than "shove it under the rug." The issues and solutions can be a resource that evolves into career options, community collaborations, volunteer endeavors, a resource directory and more.

Fashion Show- Holding the World, Working Together in
"Signs of the Time." Photo by Patrice Ward

Note: As you read, you will discover there are **words that have the same meanings:**

- The author, I, LaVerne McLeod
- Black, black, African Americans
- White, white, Caucasian, Anglo
- Latino, Latinx, Mexican, Hispanic
- BIPOC-Black, Indigenous, Persons of Color
- Earth, earth, planet, Planet, Mother Earth

There are **frequently used acronyms** in the reading as well:

- IPCC, Intergovernmental Panel on Climate Change
- USDA, United States Department of Agriculture
- NOAA, National Oceanic and Atmospheric Administration
- NASA, National Aeronautics and Space Administration
- PEW, a family name denoting Pew Research Center created and funded by the PEW Charitable Trust
- CDC - Centers for Disease Control & Prevention
- EPA, Environmental Protection Agency

Bell Tower

Chapter 1

Wake-Up Calls

As we move through (and some have moved on through) the deadly viral pandemic and divisive times, we have yet another story to be birthed. This new story involves our planet and its people and continues to unfold.

The Crossroads of Social and Climate Justice may not include every climate issue nor every social justice issue that exists. In this particular chapter, the story is more concerned with the root of these issues as a starting point. The late Jamaican activist, Marcus Garvey, said "people without the knowledge of their past history, origin and culture is like a tree without roots."

Rooted in our history we may wonder how did mankind go astray and decide to take, pillage and destroy other humans? This sad separation from connectiveness came a long time ago and still prevails today. Could it be sinister, competitive, power-hunger and greed-thirsty desires that made mankind callous and heartless? Would that same callous attitude have anything to do with what we face today?

Some humans do not want to hear about such subjects as climate change. I sometimes call it 'climate construction' for it's being built upon every day. Nor do they want to hear about race relations and would rather invalidate the issues before exploring them. If your curiosity to read this book is based on overturning and negating its validity, please take a deep breath and absorb a quote from a former enslaved person, Frederick Douglas: *"No man can put a chain about the ankle of his fellow man without at last finding the other end fastened about his own neck."* 1

Clearly, the aftereffects of this yoke that links to shame, guilt and remorse has been a dis-comforting fact that is noted in many diversity and inclusion workshops. It is also the fear of being connected to someone different than oneself that has provoked the struggle to nullify and keep a division. Here is what one enlightened white person, Marion Van Namen, permitted me to say about a social media comment she wrote: "Don't let shame and guilt stand in your way. Lives are on the line every single day. I just wanted to share openly because I remember the first time, I realized that my kids (two adopted black kids) were digesting the pain of their enslaved ancestors. My jaw dropped; really, are those leftovers of slavery?"

All of this has perpetuated too long as a world issue- the division of healing the Earth we live on and the people that live on it. Let's not forget to mention creatures and dissimilar forms of life that exist as well. Mankind has been so inclined to ignore things as a society unless it is of monetary benefit and calls it survival. As a result, our planet has suffered and its people in so many demoralizing ways. When I speak of people, I mean the poor, the meek and especially those of color.

The root of the issues in *The Crossroads of Social and Climate Justice* can be explained from some historical examples. This reflection is given as these topics occur simultaneously. As hard as we try to continue to polarize and divide without the reality of our connectiveness, struggles persist.

Found in a collection of quotes by famous people, remember the wisdom of Chief Seattle, the chief of 'Dkhw'Duw'Absh' and 'Suquamish' tribes of Native Americans, who lived from 1786 to

1866 and is widely regarded as one of the most important figures among Native Americans for his efforts to accommodate the white settlers in the United States. In 1855, here is what Chief Seattle said that is applicable to today: *"Humankind has not woven the web of life. We are but one thread within it. Whatever we do to the web, we do to ourselves. All things are bound together. All things connect."* 2

Some historical flashes rooted in history:

A. Going further back into the root of climate and its people, let's start with the Middle Ages in 15th century Europe. Here we go back into the root of climate reconstruction and its vulnerability. There was the unusual climate occurrence in the 1430's known to be the coldest years with an icy winter lasting through May. Fields and crops became frozen, sheep and other livestock died, people were starving as the food supply was almost non-existent as grains and agricultural loss added to the crisis. This impacted the rich and the poor people and drove up the cost of what could be raised or found. According to Dr. Chantal Camenisch, a history climatologist who studied historical archives at the University of Bern in Switzerland, Scotland's 1432-1433 winter was so cold that people used fire to melt wine in frozen bottles before drinking it. 3

The general health of the people with mal-nourished bodies propelled to weak immune systems. The breakout of a contagious epidemic, The Bubonic Plague or Black Death, in the Middle Ages accelerated as did the mortality rate. Guess who could purchase some food items, when available? To top it off, as with most disasters, what or whom can be blamed or be held accountable for this disaster? "In the context of the crisis, minorities were blamed for harsh climatic conditions, rising food prices, famine and plague." 4

Undeniably, this unusual frozen spell could have been caused by the random result of natural changes inherent to the climate system such as volcanic linkage or changes in solar activity at the time of this grave climate construction.

B. On another side of the globe, there had been the arrival of Christopher Columbus in the Americas. Even though there were Indigenous Nations that inhabited this land, Columbus decided to say that he discovered America the year of 1492.

In 1493, The Papal Bull, a dangerous document written in Latin the year after Columbus landed in America, was issued by Pope Alexander VI of Spain. This paper gave exclusive right to the New World that Columbus said he discovered. The Papal Bull gave power to enslave or kill those native inhabitants on the land and claim the land. 5

C. Also rooted in history is the discovery of race and the idea that was conjured to rationalize exploiting West Africans as slaves. This began in the 1450's when the Portuguese king hired Gomes de Zurara, to write a story to justify Africans as inferior people, even though the African culture was rich and sophisticated in pre-colonial times. Gomes de Zurara recorded in his chronicles that Africans were inferior due to the color of their dark skin. Thus, slave traders connected to the Portuguese crown pioneered the Atlantic trade route. They were the first Europeans to sail across to the sub-Saharan Africa (stretching from Senegal to Ethiopia) to kidnap and enslave African people. And you know who caught onto this idea as well, about the inferiority lie? 6

African men, women and children were captured and considered personal property. Their free labor created what is now the United States of America. This happened in many other countries. The institution of slavery was one of the most notorious acts on humankind, being treated worse than animals and discounted as human beings, all for personal gain.

D. With the demise of slavery's terrorized free-labor force and the scientific attention gained with the discovery and use of fossil fuels, exploiters were building nations. Attributing Andreas Latavius' theory in his 16th century writings, 7 and later after more research, that of Mikhail Lomonosov in 1757. 8

They confirmed the theory about fossilized materials. This encompasses the remains of dead plants by exposure to heat and pressure in the Earth's crust over millions of years. Thus, mankind started its exploitations of the earth's energy sources. Coal, the black chalky rock, produced lots of heat needed in the transportation industry. Boilers were invented and coal-fired engines pulled heavy trains and powered steamboats. As its usage expanded, excessive mining happened that has created to this current day, a planetary crisis.

The outbreak of mining coal in the 1600's continued with its usage on a grand scale. One might conclude that this is either, the stepping stone towards industrialization or a triggering point that calls for today's usage of renewable energy.

Furthermore, did people stop to listen to themselves when they blamed poor folks for a plague? Was there any listening and any moralistic revolt when legal papers were written up to kill and kidnap people and steal land? Did anyone care about the treatment of captured African men, women and children when tortured to work under force and duress?

It seems that folks were and are not awakened by anything that happened unless it is a threat to what they have materially or a threat to their benefits of power being taken away. Take Paul Revere on his and America's midnight ride. A great clamor went out to the militia in that April of 1775. The British were coming. The same metaphoric clamor is again upon us. 9

The writing in the US Constitution excluded the rights of an enslaved African as a free man. Then came more suppressive laws guaranteed to keep those of African descent in a suppressive world. Throughout the 19th century, this persisted with frightening laws and vigilante occurrence of terror to African Americans.

E. In 1865 after the American Civil War, there was the unlawful selling of people back into slavery. There were folks in the South determined to keep things exactly as they were during the heyday of slavery. Even though the enslaving of African Americans were illegal, slavery persisted in many other ways. There was punishment in convict camps, sharecropping debt and fake apprenticeships because no other way of life was known. 10

There is more. If an African American committed a small offense or occurrence that was considered unlawful, they were arrested and placed back into slavery or prison camp and put in chain gangs and work camps on farms, mines and quarries. Any excuse created an arrest such as using obscene language, selling cotton after sunset or using a bad word. The southerners made into law "The Black Codes" 11 and "Sundown Laws." 12

Further, the suppressive efforts were a form of slavery by leasing prisoners called the Convict Lease 1880-1926. 13 Jim Crow and Segregation customs were also enforced. These types of legal suppressions are still happening today. 14

F. Moving towards current times and rooted in climate justice are the alarming noises of highly researched, verifiable, and profound entities that did not wake up the world until the Viral Pandemic of 2019. This brought disparities to the forefront as persons of color suffered the most.

G. There was the James Hansen report, the Intergovernmental Panel on Climate Change (IPCC), 350.Org's Announcement, One Earth Plan and the Climate Strike of 2019, to mention a few unnerving signals. Attention to these alarms hinge upon action so the human species can survive. Thus, the planet and its people are intertwined for survival as one entity.

James Hansen, an American physicist, currently Program Director of Climate Science, Awareness & Solutions of the Earth Institute at Columbia University, gave a speech June 23, 1988 that rocked the world. Yes, he was ringing a bell for Mother Earth by telling the climate story in a speech to awaken everyone. This speech was given when he was working for NASA's Goddard Institute for Space Studies delivered to a panel of the U.S. Senate Committee on Energy & Natural Resources at the Capitol in Washington, D.C. This formal address may be one of the ringing bells to have started the revolution to save the earth. Hansen's report reflects the fact that human activity caused the warming climate due to high levels of CO_2 in the atmosphere. He describes consequences such as ice melting, sea level rise, droughts, extreme weather and superstorms, and even large-scale die-offs of life and ecosystems, such as coral reefs. 15

Holding the mirror to climate justice, it was seen that these efforts were not taken seriously by the Senate after James Hansen's report. Looking through the social justice lens, this denial affects the lives of individuals who have few to zero resources.

H. Social injustice like climate injustice started a long time ago so we can start at various points in history to find it. Let's look at another flash period. Topeka, Kansas, 1954, when Oliver Brown, an African American minster, wanted his daughter, Linda Brown, to attend a public school of all white students. The school was closer to his home and safer than having her walk across railroad tracks. The school denying her to enrollment created a landmark case, Brown vs Board of Education. The Supreme Court ruled segregation in public schools unconstitutional. 16

I. In 1955, the brutal murder of Emmett Till, a 14-year old African American kid was tortured, shot and dumped into the Tallahatchie River for whistling at a white woman. 17 I experienced a re-enactment of the funeral of Emmett Till and surrounding facts of his brutal murder at Smithsonian's African American Museum of History and Cultures in February 2020, before the viral pandemic shutdown. This among other atrocities upon African Americans displayed through artifacts, visual, audiovisual and written displays was painful to see in the Museum, yet was great for the world to visit and know the true history story of African Americans.

J. Only a year after Emmett Till's brutal murder, Rosa Parks, in Montgomery Alabama, refused to give up her seat at the front of the "colored section" of a bus to a white passenger, after the white section of the bus became full. She would not move as she was tired. Rosa Parks was also active in the Civil Rights Movement and was elected Secretary of the NAACP in Montgomery, Alabama at the end of 1943. She was arrested for refusing to obey a bus driver to sit in the colored section. Her actions led to a Montgomery boycott of buses by African Americans. Later, a Supreme

Court upheld a judgement that it was unconstitutional to segregate seats on buses. 18

K. Then the leadership of the Reverend Dr. Martin Luther King, Jr. with his non-violence stance emerged. As we may know, Dr. King became the hero of the Civil Rights Movement with the help of both Black and White citizens. 19

Other African American leaders, such as the belated John Lewis, US House of Representatives for Georgia, joined in to work out plans to create changes for Blacks with nonviolence. As he dedicated his life to advancing human rights, he spoke to current day voting suppression in 2019 effecting Black voters: *"The vote is precious. It is almost sacred. It is the most powerful non-violent tool we have in a democracy."* His famous quote: "

"Do not get lost in a sea of despair. Be hopeful, be optimistic. Our struggle is not the struggle of a day, a week, a month, or a year, it is the struggle of a lifetime. Never, ever be afraid to make some noise and get in good trouble, necessary trouble." 20

And we know what the belated (1924-1987) African American playwright and author, James Baldwin, has to say along these lines. "Not everything that is faced can be changed, but nothing can be changed until it is faced."

This serves as another good reason that I am still pursuing the conversation to elicit change.

Violent deaths occurred as a result of vigilantes who did not want African Americans to advance. There was a church bombing that killed African American children attending Sunday School. 21 The turning of police dogs on peaceful marchers 22 and the eventual assassination of Dr. King happened. 23

Blacks did not give up, rather they continued to press for equality, jobs and the right to vote without restrictions. The Equal Employment Opportunity Act was passed laying the groundwork for affirmative action. Pulitzer Prize winners, Alex Haley, 24 Alice Walker 25 and August Wilson 26 emerged as were other African American achievers, yet the United States was still far from being a just or equitable nation.

L. Turning the mirror back into view on the climate side of things, we can see that not enough political heads turned in the positive directions towards healing the planet that also affects its people. This is evidenced from reports or leading stories about the climate. The reports refer to the IPCC known as Intergovernmental Panel on Climate Change. The IPCC is the leading international body for the assessment of climate conditions. It was established by the United Nations Environment Program (UNEP) and the World Meteorological Organization (WMO) in 1988. Also, the IPCC provides the world with a clear scientific view on the current state of knowledge in climate change and its potential environmental and socio-economic impacts. In the same year, the UN General Assembly endorsed the action by WMO and UNEP in jointly establishing the IPCC.

The IPCC, under the supervision of the United Nations (UN), reviews and assesses the most recent scientific, technical and socio-economic information produced worldwide relevant to the understanding of climate change. It does not conduct any research nor does it monitor climate related data or parameters. Thousands of scientists from all over the world contribute to the work of the IPCC on a voluntary basis. [27]

I interviewed Dr. George Somero and learned that Dr. Somero, a local California scientist, contributes his expertise to the IPCC. He is the David and Lucille Packard Professor of Marine Science Emeritus and currently works at Stanford University's Hopkins Marine Station in Pacific Grove, CA where the interview was held.

Dr. Somero talked about the IPCC as a solid foundation for climate findings.

He, along with about 25 other ocean scientists, further reviews and edits the "portion on the ocean." A major task of such editing is to help clarify the points made by many other scientists from many nations, such that the final document is readily understood by diverse readers across the globe. He indicates that preparing an IPCC report is a several-year process that begins with diverse lines of research. It ends with the final polishing of a long and well-documented Report. Somero emphasizes that the oceans face numerous major problems from global change, including sea level rise, increasing acidity and increases in temperature that threaten marine ecosystems. Many US cities, notably Miami, are heavily threatened by sea level rise.

An IPCC Report as Somero puts it, is *"absolutely"* a reputable valuable source about global warming. *"And which isn't to say you won't make mistakes particularly when you are relatively working with incomplete data. For example, back in 2007, the rate at which sea level is likely to rise, wasn't really clear. There wasn't adequate information on the decay of the Greenland ice sheet. They now have this information and can make a more accurate prediction about sea level rising. The predictions are essentially the same and now more precise as: instead of saying that temperatures are likely to go up one degree Celsius, they are saying it looks like it is going to be between two and four degrees Celsius. So, the IPCC is refining their predictions. One degree Celsius is relatively 1.8 Fahrenheit. Therefore, when you hear IPCC reports for example, most climate temperature data, most are reflected on the Celsius scale. Fahrenheit is largely used only in the United States. When you hear it is gone up only 1 degree, that means it has gone up 1.8 degrees Fahrenheit. It is not quite a 2 to 1 thing but the sound of 1 degree seems to lessen the blow that is almost 2 degrees."* 28

It is interesting to note that the IPCC assessment that was available at the time of the interview showed interesting results of a couple of points I will emphasize from sections of the assessment report.

A-1. *Human activities are estimated to have caused approximately 1.0 degrees Celsius of global warming above pre-industrial levels with a likely range of*

0.8^{0}C to1.2 degrees Celsius. Global warming is likely to reach 1.5 degrees Celsius between 2030 and 2052 if it continues to increase at the current rate.

B.4.2. *Global warming of 1.5°C is projected to shift the ranges of many marine species to higher latitudes as well as increase the amount of damage to many ecosystems.* 29

The IPCC has definitely rung the bell for Mother Earth and so has the role of Al Gore, a political climate crusader. In 2007, the

IPCC and U.S. Vice-President Al Gore were jointly awarded the Nobel Peace Prize "for their efforts to build up and disseminate greater knowledge about man-made climate change." [30]

M. Have we not paid attention to the IPCC that continually rings bells what we close our ears to? Are we going to let our planet and its people be exploited and eventually wilt with increased heat? Can we awaken to what Bill McKibben, a significant environmental activist, leading journalist, author and Middlebury College Scholar in Environmental Studies has to offer? McKibben along with university friends founded an organization called 350.org in 2008. It is the first planet-wide, grassroots climate change movement. As an author, McKibben's *The End of Nature* (1989) was the first book for a general audience about climate change. In 2002, he wrote a review to explain *The End of Nature*. Here is one excerpt:

He says that the key environmental fact of our time is *"the contrast between the pace at which the physical world is changing and the pace at which the human society is reacting."* He goes on to say that *"man can no longer think of himself as a species tossed about by larger forces – now we are those larger forces. Hurricanes and thunderstorms and tornadoes become not acts of God but acts of man."* [31]

With the formation of 350.org, the goal is to end the use of fossil fuels and transition to renewable energy by building this grassroots movement globally. McKibben asked world leaders to address climate change and to reduce levels from 400 parts per million to 350 parts per million. [32]

Hence, the name of the organization.

So where did these numbers come from?

It came from Hansen in the paper he wrote in 2007 that 350 parts-per-million (ppm) of CO_2 in the atmosphere is a safe upper limit to avoid a tipping point. i.e. when things go over their limit and tip over and cannot be reversed. [33]

An example would be the popular reference: tipping over a glass of water and knowing that a lot of water spilled out even though you quickly stood it upright again and that water cannot be replaced back into the glass.

As a journalist, McKibben writes for the *New Yorker, National Geographic,* and *Rolling Stone.* In his 2012 *Rolling Stone* article, he explains "Global Warming's Terrifying New Math." 34

To summarize this particular article, there are three simple numbers that add up to global catastrophe:

The first number is 2⁰ Celsius. This is the ceiling number that we can't afford to go beyond if we wish to avoid the worst consequences of climate change.

The second number is 565 gigatons. It means that if we throw more than 565 gigatons of carbon dioxide into the air between now and 2050, then we will almost certain exceed the 2^0 Celsius limits.

The third number is 2795 gigatons. This represents the amount of carbon dioxide that would be emitted if we burned up the planet's current reserves of fossil fuels. This refers to the oil, gas and coal assets that presently exists in the ground.

McKibben states: *"we remain in denial about the peril that human civilization is in."* 35

N. To continue with a full-length view of the mirror, I decided during the course of writing this book, to attend the Bioneers Conference in San Rafael, CA in 2018 and get yet another taste of the climate movement that affects our planet's people. There was a very interesting report from Ms. Justin Winters at this conference through her speech in answering this question: 36

"How can humans halt the ever-more-intimidating threat of global climate change before catastrophic results?

Ms. Winters, Executive Director of the Leonardo DiCaprio Foundation (LDF) at that time, unveiled what the Leonardo DiCaprio Foundation has sought to address, the One Earth Plan. This plan aims to avert a climate crisis and protect our biosphere.

The original focus had been on developing a global plan to protect nature. Ms. Winters indicated that *"you can't protect nature without also addressing climate change and making sure that humans have enough food and water to survive.*

She further disclosed that" the *LDF commissioned research*

from over 20 top climate, energy, conservation and agriculture scientists around the world and arrived at a three-pillar plan, a vision for the planet that LDF calls it One Earth, where protecting nature plays a critical role."

As she continued to speak, she stated that, *"we can ensure that our global temperature doesn't go above 1.5 degrees Celsius if we achieve three goals by 2050:*

1. Transition our energy systems to 100 percent renewable energy.

2. Protect, connect and restore 50 percent of our natural ecosystems on land and sea.

3. Transform our agriculture systems to regenerative and carbon negative agriculture practices (or regenerate 50 percent of global working lands to build healthy soil for food and fiber production by 2050).

It's an ambitious plan, but given what's at stake – what other option is there? This is the moment when we are – as a movement and as a people – being asked to step up."

O. And then the sounds got louder with the world first hearing a 16-year-older, at the time, brave young, Greta Thunberg. It doesn't get any clearer than this and more urgent to focus on our climate issues as we look at the younger generation now taking the helm.

Greta Thunberg speaks clearly and intensely as her speaking ability has made an impact on her listeners all over the world. For example:

Starting in August of 2018, Greta Thunberg started a school strike for the climate outside the Swedish Parliament. This event currently involved over 100,000 schoolchildren. The movement is now called "Fridays For Future. "

Thunberg has spoken at climate rallies in Stockholm, Helsinki, Brussels and London. In December 2018, Greta Thunberg attended the United Nations COP24 in Katowice, Poland. She addressed the Secretary-General and made a comprehensive speech that went viral and was shared many million times around the world. 37

In January 21, 2020 Greta was invited to the World Economic

Forum in Davos, Switzerland where her speaking again made a worldwide impact. The Forum was filled with celebrities, politicians and business leaders. Her speaking was refined and intentional, combined with her stoic manner. The attendees may have been shocked with this excerpt: as she spoke:

"We demand at this year's World Economic Forum, participants from all companies, banks, institutions and governments immediately halt all investments in fossil fuel exploration and extraction.

Immediately end all fossil fuel subsidies.

And immediately and completely divest from fossil fuels.

We don't want these things done by 2050, 2030 or even 2021. We want this done now.

It may seem like we're asking for a lot. And you will of course say that we are naïve. But this is just the very minimum amount of effort that is needed to start the rapid sustainable transition." 38

To capture more worldly attention to the climate issue, Ms. Thunberg's voice was heard as school strikes with youth and adults culminated into a *Global Week for Future,* September 20-27, 2019 Global Climate Strike. 39

Millions of strikers protested globally. 40 The lead figures were Future Coalition groups. As an environmentalist and social justice activist, I found myself, along with friends, protesting in the Global Climate Strike Day in Monterey, California. I am pictured with my "Equity for All and the Planet" sign to represent social justice as well as climate justice. Accompanied by friends, Kay Cline, with her "Earth Matters" sign. In the bottom photo another friend, Kristin Ramsden and her generational family members amongst many concerned citizens, alerted vehicle passengers on Del Monte Street of Monterey, CA about the climate crisis.

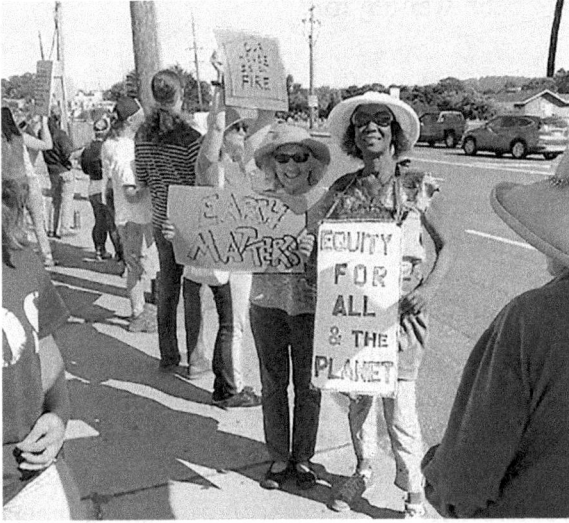

Climate Strike in Monterey, CA.

Climate Strike in Monterey, CA.

Another speech of inspiration that perhaps enticed the strike is "The Disarming Case to Act Right Now on Climate." 41

Well, now are we listening? Is the viral pandemic COVID-19

what we have been waiting for?

P. If you haven't been listening to the bells that were rung in this chapter, you certainly stopped for the pandemic as you understood that it was a life and death situation. So, what did we learn about climate change with COVID-19 when we stopped most automobiles, factory production, transportation in the air and were sheltered in-place?

We learned, by some unfortunate, yet wonderful miracle, we could see that Mother Earth cleansed herself. Now, we learned that when we as humans get out of the way, Earth takes care of itself. Blowing toxic smoke and greenhouse gases into her atmosphere, spewing greenhouse gases on Earth's landscape with automobiles and other vehicles; from the air with airplane travels and in the waters with cruise vessels that festered with confirmed cases of COVID-19 throughout its manifest.

Can one say that our Earth Mother spread a dis-ease amongst us due to our exploitation of her resources? It surely is a coincidence. Maybe there will be an answer to what caused the pandemic or not. The intent is not to focus on the causes of pandemic. The fact that we really don't know what or what combination of other variants are involved, there are just a few interesting concepts to mention. After all, it did happen on this planet and affected our vulnerable elderly and mostly persons-of-color with deadly consequences.

The concept taken by the framework of One Health recognized by the Center for Disease Control, the World Health Organization, and governments and organizations around the world, considers man's relationship with animals and exotic pets to disease. Interestingly, others chime in on this possibility.

Rachael Bale's National Geographic Animal Stories Executive Editor and other authors of animal stories, correlate sick animals to sick humans as a possible cause for the global COVID-19 virus. Bale indicates that the pandemic is a result of humankind's destruction of the planet. As her fellow writers have also indicated:

"Rampant deforestation, uncontrolled expansion of agriculture, intensive farming, mining and infrastructure development, as well as the exploitation of wild species have created a 'perfect storm' for the spillover of diseases from wildlife to people." 42

In any event, no matter what caused the COVID-19 pandemic, let's look at what happened when mankind had no choice but to see nature's recovery:

A rare sight of dolphins swimming in the Ganges near the shore. This is done without the disturbance of fishermen in a region in the Bosporus Strait between Europe and Asia. This means a healthier maritime ecosystem took place without man. 43

There's less air pollution accordingly to Marina Koren, a writer for the Atlantic. This is evidenced by some valid facts from the World Health Organization and Marshall Burke, of Stanford University:

"As cities and, in some cases, entire nations weather the pandemic under lockdown, Earth-observing satellites have detected a significant decrease in the concentration of a common air pollutant, nitrogen dioxide, which enters the atmosphere through emissions from cars, trucks, buses, and power plants. The drop, observed in China 44 *with stringent social-distancing measures on the ground also occurring in Europe.* 45

Air pollution can seriously damage human health, and the World Health Organization estimates *that conditions stemming from exposure to ambient pollution—including stroke, heart disease, and respiratory illnesses—kill about 4.2 million people a year.* 46

According to an analysis by Marshall Burke, a professor in Stanford's Earth-system science department, 47 *a pandemic-related reduction in particulate matter in the atmosphere—the deadliest form of air pollution—likely saved the lives of 4,000 young children* and 73,000 elderly adults in China over two months this year (2020)." 48

Less traffic in the USA, less CO_2 emissions in the atmosphere. Chelsea Harvey writer for the E&E News Environment states: *"The spreading virus has caused a dip in global greenhouse gas emissions. Reasons include a temporary blow to industrial activities in China, falling demand for oil and a decline in air travel. In China, the world's largest carbon emitter, experts estimate that emissions over the past month (Feb 2020) have been about 25% lower than normal."* 49

COVID-19 had devastating effects upon all the world. With less shopping, traveling to work and businesses and schools closing during shelter-in-place, public transportation all but came to a screeching halt. This has affected CO_2 emissions and therefore there was less noise, less fog and fresher air.

Now, the question is: Will mankind make every effort to change habits that create less polluted air and greenhouse gases into the environment after the lessons learned from the quarantine?

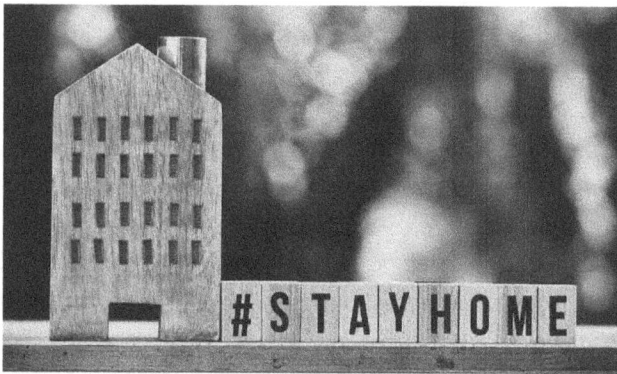

Call at Home Stay

And what about those who couldn't comply to such postings as "Stay Home, Save Lives?" Someone had to take the brunt of the catastrophe as the pawns. Speaking about those out front, I mean the ones who had to work in order to continually have a roof over their heads and could feed their families. Could we make this a different world where we all have equity? Think about it since we have witnessed deadly blows of end-of-life to many frontline workers, the vulnerable and to people of color.

Q. It's as if the universe says this isn't enough, guess what? The social justice movement got jolted again. This time it was the public lynching of George Floyd an African American man who was intentionally murdered by a white police officer. That murderer showed off to his fellow police officers who just watched and did nothing. This happened right after the quarantine. 50

Locked up with a virus surrounding us and then clogged and shocked for what has been going on for centuries and only then did the world wake-up. Blinders have been removed by so many and the work of racial healing is on the rise. This prompted me, in the midst of the pandemic, to create two series of Bridge Building to Equity - Racial Pandemic Webinars and Workshops (2020 and 2021). I was forced to facilitate online as I was driven to do something in the moment. Others took similar actions while thousands took place in passionate marches across the globe and some folks later dropped the ball. It is to our best interest to continue with racial and planetary healing. Let's do so together even if you are afraid. Take to heart what the late Dr. Martin Luther King, Jr. said: *"Take the first step in faith. You don't have to see the whole staircase. Just take the first step."*

Were you sheltered in place during COVID-19 or did you do essential work?

The Elemental Mirror of Air

Air/Heaven

What's in the Air?
What's in the Air that we breathe?
What's in the heated environment that truthfully airs justice?
What's in the environment that deletes equity for struggling people or people-of-color in our world?

What solution(s) can we act on to create social justice and climate justice where there's harmony for all people and the planet?

Air reminds us that there is much more to this world than what we can see, and also teaches that we must grow and change as the world does.

Even if we don't know all the answers right now, let's open up to some of these issues and solutions presented in the next eight chapters of this section. Please commit to peacefully act on one or some of them to help create a better world!

Two Girls Awaiting Help-Hurricane Katrina.

Chapter 2

The Risk Where You Live

Nothing is more important than our home. It's the place where we lay our heads each night, where we enjoy reading books, cozy up to the fireplace, unwind after a cumbersome day of work or projects, where we contemplate our dreams. Hopefully, we learned the value of home when we were sheltered in place at the outset of the COVID-19 pandemic.

This may not be the case for everyone. For some there is the uprooting due to war, leaving one's home for a better opportunity to live and feel safe, catastrophic weather conditions where homes are destroyed and other reasons.

With rising temperatures and extreme weather comes a host of issues affecting vulnerable-older folks and persons of color. I was struck with grief and anger learning about the social injustice that came about with Hurricane Katrina in 2005. Neighborhoods were flooded and families sat on rooftops of their overflowed homes for days with no food and water, waiting to be rescued.

According to a climate research group called Moody's Environmental Social Governance (ESG Solutions) formerly known as Four Twenty-Seven, there are key findings that affect where one lives. Here are some risks where you live on a climate change or climate construction level. As these are discussed, there will be reflections on how social justice is affected with where one lives. 1

The first key point reflects the rising sea level. It is explained as an effect of global warming. So, when the waters become warmer, there is expansion that results in oceans rising worldwide. Mostly, however, sea level rise reflects more water entering the seas from melting ice (glaciers, ice shelves and ice sheets). Among the areas affected in the USA are mid-Atlantic states, particularly New Jersey, Virginia, North Carolina and Florida. These domains have the highest exposure to coastal flooding in the United States, with the Pacific Northwest also highly exposed in several of their coastal cities and counties.

We see what happened in states such as New Jersey where sea level is massively rising. The expansive water or "thermal expansion" is responsible for about 40% of global sea level rise in the last 25 years or so according to Rutgers New Jersey Climate Change Resource Center. They also speak to the melting of glaciers and ice sheets being another major contributor to rising oceans. For example, there is evidence that ice loss in Greenland and Antarctica is speeding up therefore accelerating more sea level rise. The loss of glaciers and ice sheets currently accounts for about 45% of global sea level rise. In the case of New Jersey's rising sea level, there are other issues involved that are not related to thermal expansion. Those issues are a sinking mid-Atlantic region while land to the north once covered by Ice Age glaciers rises up.

Another issue is the pumping of large amounts of water from aquifers that adds to the sinking of New Jersey's coastline. A lot is lost, resulting in degraded coastal ecosystems; seawater entering into the woodland areas causing leafless branches and fallen trees; the washing away of once attractive beaches; the drowning of marshes that serve as a marine nursery and a natural buffer against flooding. 2

On the social justice side of **rising sea levels**, in states and countries where rising sea level is eminent, it is the vulnerable and mostly BIPOC populations that are affected with extreme recovery issues. This includes increased health risk of diseases and infestations with already compromised immune systems, food and clean water scarcity and being overburdened by caregiving those in need. The system of gentrification whereby real estate investors buying out inner-city and poor neighborhoods to sell for a high price to wealthy persons, results in eventually pressuring out the once poor neighborhood. Black, Indigenous and People of Color (BIPOC) populations are pushed to affordable areas of living where more frequent and harmful climate related disasters occur.

We look into the mirror and we also see Van Jones, founder of Green for All, is quoted as saying "the traditional environmental movement has a diversity problem. And for the environmentalism as a whole to succeed, that needs to change." He further indicates that voices are ignored due to small vs large scale environmental movements funding. "Diversifying the donor lists of foundations that usually give to environmental groups would help black Americans in particular make voices heard in the environmental movement." 3

According to Students 4 Change, indigenous voices are often excluded from decision-making processes, despite their wisdom and experience dealing with the burden of climate change.

Another key issue that Moody's ESG Solutions points out that involves the risk where we live is **cyclones/hurricanes**. A cyclone is an atmospheric system where the air is traveling rapidly and circulating around a point where the pressure is low and after the passage of time is accompanied by destructive weather. A hurricane is a cyclone that reaches or exceeds 74 mph. 4

Emphasized with Moody's ESG Solutions is that the majority of cyclone risk in the United States is concentrated in the Southeast, given its geographic proximity to the Gulf of Mexico and the tropical Atlantic Ocean. The coastal Mid-Atlantic and Northeast are also exposed to cyclones, but they tend to be less frequent than in the Southeast and somewhat weaker on average after interacting with land or cooler ocean waters.

There are several hurricanes that could be mentioned, it was Katrina in New Orleans that stuck in my head. Some people of color who needed to be rescued and have shelter were disregarded. The neighborhood where poor Black folks lived sat below sea level. This meant that they were doomed should the levee break. Even though the Mayor, Ray Nagin, at the time, issued the city's first-ever mandatory evacuation order the day before Katrina hit, folks living in those dooming neighborhoods had no place to go to or couldn't get there.

"The Mayor declared that the Superdome, a stadium located on relatively high ground near downtown, would serve as a "shelter of last resort" for people who could not leave the city. (For example, some 112,000 of New Orleans' nearly 500,000 people did not have access to a car.) By nightfall, almost 80 percent of the city's population had evacuated. Some 10,000 had sought shelter in the Superdome, while tens of thousands of others chose to wait out the storm at home." Some managed to make it to the Superdome while others who got there found that it was full and locked.

To exacerbate the issue, many scrambling for shelter were turned around by police with guns when trying to enter into neighborhoods where rich white folks lived on higher ground. They had walked over the Crescent City Connection bridge to the nearby suburb of Gretna. The US President at the time George W. Bush did not respond quick enough to this grave emergency. Black folks were literally sitting on their rooftops, starving and thirsty, for several days with their families in hopes of being rescued while their home was under water. This lack of interest in saving these humans was not what the I thought would happen in the United States of America with its vast resources for emergency rescues. 5

Reflecting at this very moment, consider where you live. There are climate risks where we all live. Earthquakes, fires, hurricanes etc. does not distinguish rich from poor and a virus definitely doesn't. One's resources can help one choose where the calamities may not happen as frequently, yet can't be avoided all together. There are those who have no choice, according to the National Oceanic and Atmospheric Administration (NOAA), except to feel extreme heat in their neighborhood. In 2020 NOAA reports that it was the Earth's second-hottest year of the past 140 years, indicating a climate change.

As we focus on climate and related terminology, let's be sure that we make distinctions with the definitions. According to National Aeronautics Space Administration (NASA):

Weather: Weather is the short-term changes we see in temperature, clouds, precipitation, humidity and wind in a region or a city. Example: In the morning the weather may be cloudy and cool and by afternoon it may be sunny and warm.

Climate: The climate of a region or city is its weather averaged over many years. This is usually different for different seasons. Example: a region or city may tend to be warm and humid during summer. But it may tend to be cold and snowy during winter.

Climate Change: refers to any long-term change in Earth's climate, or in the climate of a region or city. This includes warming, cooling and changes besides temperature. Note that the author may sometimes refer to climate change as climate construction.

Global Warming: refers to the long-term increase in Earth's average temperature. [6]

What if you lived where there is **extreme rainfall**? This is another grave issue, particularly in the Midwest. This area is particularly exposed to increasing amounts of flooding because of changing rainfall patterns. Recent advancements show extreme rainfall to be the main cause of recent floods and not agricultural practices.

Take for example Hurricane Harvey that in August of 2017 that started as extreme rainfall that evolved into a hurricane and created mass flooding. Harvey was a devastating category 4 hurricane, where the wind speed was sustained for 1-minute at 130-15 mph or 209-251 kilometers per hour. This made the swirling turbulence move from water and onto land and landfall. In Texas and Louisiana in August 2017 catastrophic flooding caused more than 100 deaths. [7]

You have heard this story as previously mentioned about BI-POC and recovery from disasters. Keep holding the mirror and look into what also happened after Hurricane Harvey that created mass flooding. BIPOC communities located in areas along the US-Mexican border in Texas are prone to flooding with a lack of sufficient wastewater infrastructure.

Many wealthy Texans had the means to evacuate their homes while poor and disabled residents were forced to brave the storm. Undocumented immigrants were pressured to stay where they were for fear of being taken into custody with Border Patrol. This Patrol was not merciful and declined to suspend its checkpoints even during Harvey's disaster. 8

Another factor affecting where people live is **extreme heat or heat stress,** a major discomfort. It is why the focus is on cooling the earth down from its intense worldwide warming. The key finding for the United States is the highest heat stress scores tend to be centered in the Southeast and Midwest, concentrated in Missouri and western Illinois and fanning out to the Great Plains, Mississippi River Basin, and Florida.

According to the Centers for Disease Control and Prevention (CDC), "the Midwest is projected to have the largest increase in extreme temperature-related premature deaths. Northern midwestern communities and vulnerable populations that historically have not experienced high temperatures may be at risk for heat-related disease and death (dehydration and heatstroke)." 9

"Disproportionately represented by people identifying as Black, African American, Hispanic, or Latino outdoor workers are particularly at risk as climate change makes dangerously hot days more frequent and intense," says Union of Concerned Scientists (UCS). 10

Key watersheds for agricultural production such as the Central Valley aquifer system in California and the Ogallala Aquifer in the Great Plains are highly exposed to **water stress**. This affects where one lives and works. The areas of Bakersfield, Delano, and Visalia, CA along the Central Valley Aquifer are among the ten cities most exposed to water stress. Municipalities along the Ogallala Aquifer in the Great Plains also rely heavily on agriculture and are among the most exposed to water stress.

Fires, although not included in Moody's, are an eminent issue since many persons live in the wildland urban interface where the indigenous practice of controlled burning is not used. Therefore, the fuel of fallen dead trees and debris on forest and wooded floors has the potential to create intense fires with all the elements to make it happen: oxygen, heat and fuel.

Although fires will be covered in Chapter 16, it is mentioned here as this is a critical risk for where folks live. The CDC indicates that long periods of record high temperatures are associated with droughts that contribute to dry conditions and drive wildfires in some areas. As far as health is concerned, exposure from wildfire smoke increases respiratory and cardiovascular hospitalizations; emergency department visits; medication dispensations for asthma, bronchitis, chest pain, chronic obstructive pulmonary disease (commonly known by its acronym, COPD), and respiratory infections; and medical visits for lung illnesses. [11]

Considering the issues at hand with natural disaster happening, where will we live in 20 years, where we can feel safe from disasters? Or is there such a thing?

The challenges continue especially for those around the globe that live along coastlines. It brings about the question of migration due to coastal washouts via storms, tsunamis or other weather catastrophes. Some people are on the move or know they will be moving back and forth from a home disaster and back again when the catastrophe has subsided. In other words, migration has become a coping skill for living with degradations of the environment.

This migration trend is happening in the Philippines and is due to repeated typhoons. In Fiji, the trend is move away to a permanent residence inland to avoid the inevitable as the government is advancing this effort.

Let's suppose that you live in Mohammadpur, India. This coastal village was built on top of the Meghna River delta located at the mouth of the Bay of Bengal. Well, deltas are a kind of landmass formed when sediment carried by rivers is deposited where that river meets a larger body of water; their borders gradually change as rivers bring more sediment in and storms wash sediment away. It is noted that river deltas are fertile ecosystems capable of supporting abundant agriculture and marine life. With an abundance of intense cyclones having caused frequent flooding; this restrains farming and fishing. These floods also erode the coastline, allowing later storms to wipe away land altogether.

Other areas such as in the United States, existing along the coastline in Alaska, the melting permafrost is causing big chunks of ice to erode causing some Alaskans to re-settle due to these changes. If one cannot move inland as in the low-lying Pacific Island nations of Kiribati and Tuvalu it means moving to another country. Even with building mangrove forests, storms are coming at such a rapid rate it is hard to keep up with water flow into villages. This is such a challenge for coastline areas since its inhabitants want to keep their same means of lifestyle and livelihood. 12

Another social justice issue that prompts one to leave their home is an issue affecting millions of people seeking refuge. These are refugees or folks forced to leave their homes to escape war, violence, and persecution. The majority of them have become internally displaced persons, which means they have fled their homes but are still within their own countries. If only there were laws to protect migrants and refugees, it would be an easier process. According to Othering & Belonging Institute, "Across international humanitarian law, human rights law, refugee law, and other bodies of law, protections for climate-induced displaced persons forced to cross international borders are limited, piecemeal, and not legally binding. International migration following short-term disasters is only occasionally protected under humanitarian visas and state-specific measures, such as with the United States' Temporary Protected Status designation, though such protections are often provisional and not legally binding. Likewise, international migration following long-term disasters is not covered unless the provision of support by the local government (or governments) is denied on the basis of race, religion, membership of a particular social group, or political leaning." 13

Consider these injustices and what can be done about them.

ACTION

THE NEW STORY OF HEALING & SOLUTIONS

Every solution and resource births a new story for change. If we want to see changes, we must take action.

In this part of the chapter are possible solutions and resources to help inspire actions so we can have opportunities to live new or revived stories to:

A) Awaken to the consequences that are risky where people live and help reduce the greenhouse gases in the atmosphere, benefitting both people and planet.

B) Find any action item as a pathway towards a climate career, social justice career, research or volunteer project that is just waiting for you to choose from. That is, unless you have already been triggered into action.

List of possible action items to pursue:

Possible Action Items to Pursue:

1. **For help with rising sea levels**, consider the following recommendations from Rutgers University Institute of Earth, Ocean, and At mospheric Sciences via network of Rutgers Engineers Infrastructure group:

- Mobile Wall which can be raised up when sea levels rise, [preventing damages to buildings and homes and cutting off habitats for important bird and fish species with a permanent wall}.

- Porous Parking Lots

- Rain Gardens

- Harvesting Local Energy from wave, wind, and rain water which would "power the pump to turn the storm against itself."

- Reduce Greenhouse Gas Emissions

- Build Wider Foundations for a big city- So that houses and buildings that are established are not standing on the ground but on the foundations. This foundation will function that are as a weight support building on it as well as an irrigation system. Thus, the pressure on the soil surface will decrease and prevent ocean level rise.

2. **Research and Study** through various Engineer Infrastructure groups from various Universities. Here are just a few:

- Carnegie Mellon University, www.cmu.edu, (412) 268-2000, Pittsburgh, PA

- University of Dayton, www.udayton.edu, (937) 229 1000, Dayton, Ohio

- Cambridge University-United Kingdom, phone +44 1223 332600, fax+44 1223 332662, Cambridge, England

- KTH Royal Institute of Technology, Phone +46 8790 6000, fax +46 8790 6500, Stockholm, Switzerland

3. **Explore careers in Disaster Management** at various colleges and universities such as:

- Disaster Management Director or Specialist
- Materials Remover Specialist

- Fire Inspector and Investigator

- Hydrologist

- Forest and Conservation Technicians

4. **Help with social justice** by strengthening and creating safe and economically secure environments where people live- individually (fundraiser, advocate, ally, donor); take action as a family; take action with social justice organizations; get involved with legislation:

- Support BIPOC businesses

- Retrieve housing statistics – on how many African Americans and other ethnicities get loans to purchase houses and land or build a house to increase your knowledge about this matter

- Help African Americans & other BIPOC families via influencing to get education on types of loans such as becoming aware of balloon payments, etc. This information can be taught or explained through: Adult Education courses and if applicable, an elderly help center as Alliance on Aging in some counties and regions

- Employ more people of color & different ethnicities

- Support those who rally for minimum and higher wages

- Stop Redlining Practices- sell your home to a BIPOC

- Strengthen labor unions for farmers

- Strengthen the labor unions for service employees

- Offer Job trainings for Courses for re-entering from

incarceration + give an increase intern wage

- <u>Share your skill</u> with someone who wants to learn It

- <u>Overcome NIMBT</u> (Not in My Back Yard) for new affordable housing

- <u>Support planning & design</u> Integrated housing communities (No Walls)

- <u>Grant females equal pay as a male's salary</u> for the same type of job

- <u>Increase minimum wages</u> so families can at least have a decent income to survive and more so, to thrive

Let's not forget that many of the above actions will also help:
- the winged ones
- the crawlers
- the swimmers
- the hoofed and claw-toed ones
- the microscopic ones

Communication

Chapter 3

Change The Narrative

It is necessary to do our best to communicate more effectively if we expect to make positive changes by gaining trust to work with one another. After focusing on the risk of where we live, it's time to introduce some words, expressions and other sayings that lash out from our tongues. We have the ability to empower or destroy one another and our planet by what we say or declare. Dealing with certain populations in our society we can speak words that solidify injustice and hurt without even knowing it.

Our thought processes may come from genetically embedded beliefs reinforced generationally. Rather than perpetuate social injustice with our words and expressions, we can decide if we will make a change or not about certain verbiage.

Let us look at what happens as we deliver narratives by telling a story, giving a statement, account or description. As we plow through this topic, mirroring our planet and its people at the crossroads, we can also decide a response to these questions:

Should the narrative be changed? What would be the effects if it changed?

The question may be what folks that are so set in their ways, are worried about, that they are determined to not understand. By placing this topic of changing narratives into various sections, we may decide to choose words and terminology that are not hurtful to others.

Words Have Power

Words can build, confine, break and make. There are so many quotes that evidence this power:

Swami Vivekananda and others quote the evidence of word power: *"The tongue has no bones but is strong enough to break a heart. So be careful with your words."*

"Raise your word, not your voice. It is rain that grows flowers, not thunder." Rumi

"Words: So innocent and powerless as they are, as standing in a dictionary, how potent for good and evil they become in the hands of one who knows how to combine them." Nathaniel Hawthorne

"Words can inspire. And words can destroy. Choose yours well." Robin Sharma

"If we understood the power of our thoughts, we would guard them more closely. If we understood the awesome power of our words, we would prefer silence to almost anything negative. In our thoughts and words, we create our own weaknesses and our own strengths." Betty Eadie

It is too easy to speak words that deny, make excuses about

and place blame on others concerning global warming. Here are some examples:

"Global warming is not actually happening"

"I haven't noticed the change of weather around me"

"Why should I have to do anything when China is the great polluter"

"Temperatures vary naturally"

"There's insufficient scientific evidence"

In other words, climate change or global warming can create an opportunity to do something about it or can create "buzz" words for non-believers.

Enslavement and Word Control

In our own mirror we might see the existence or non-existence of a warming earth. There is a need and it is a right for everyone to feel a sense of belonging on this earth. What has held up for centuries are the false stories about people of color that prevails in one form or another today. These false stories perpetuate ideas of supremacy for some folk with a light skin color and inferiority for folk with dark skin. An insight into the superior-inferior story about race and skin color was referred to in Chapter 1 with the enslavement of Africans.

In order to gain control of enslaved persons, the use of their total language, including millions of words, were forbidden. This was, in part, to avoid escape planning or plots. Words were as the expression goes, "whipped out of you." If an enslaved person wanted to live, that person could not communicate in their language. A cat-a-nine tails could be used bringing excruciating pain. One could die under blood-letting whips. This forced-word control was a process of mass assimilation or absorption into another culture, the colonial culture of servitude.

Cultural Appropriation

Similarly, Native American Indians, sometimes called Indigenous Nations, the original landowners of what is called the United States of America, were pillaged, tortured and enslaved. They were driven off their land into reservations and scattered even more if that land became valuable to the white man. Today, the names for places and areas are taken by some businesses, sport teams and groups. Native American scholars, activists, and leaders agree that stereotypical mascots and place names are damaging. They perpetuate racist views of Indigenous people. An example is what is happening in Fresno County, California.

On the land of the Dunlap Band of Mono Indians in Fresno County, California, a continual struggle is underway to change a town's name from "Squaw Valley" to "Nuum" Valley, meaning the people's valley. Other suggested names also came forth. This endeavor is being headed by Roman Rain Tree. It was the famous ski resort in Lake Tahoe, California that changed its named from "Squaw Valley Ski Resort" to "Palisades Tahoe" which brought to the forefront the name struggles. Indigenous, locals and others took note of Squaw Creek, Squaw Peak, Squaw Hollow, Squaw Flat and other names with "squaw" knowing that now, name changes could actually happen.

What this has meant to folks who have expressed their views on the matter is that "squaw" is a derogatory name. A resident and supporter of name change is an elder member of the Dunlap Band of Mono Indians says "Squaws were worse than dogs - used and traded, and killed and shot." Supporters express the use of the "squaw" town name as divisive, oppressive epitaphs and most of all: an ugly profanity for vagina, as an enslaved one.

Even though with these reasons for a name change, further devaluation and lack of recognition continues. Some opponents say "nobody in Squaw Valley wants a name change." This implies that the Dunlap Band of Mono Native Americans were no-bodies. The struggle continues. The outlook is hopeful. One day Raman Rain Tree and supporter efforts will create change. 2

Mascot names can also be damaging. According to Stephanie Fryberg, a member of the Tulalip Tribes and professor at the University of Michigan. She says, "There's no way that the use of Natives as mascots is honoring." Could this also be true when a business is named after an Indigenous Nations group? So, what happens even when Indigenous Nations see mascots or imagery as positive? Can they still do psychological harm by damaging the self-esteem (self-confidence and respect) as well as ambitions of Indigenous Nation youth? 3

Displaying sports mascots with names and misguided cultural regalia is a way of minimizing the views and cultures of Indigenous Nations. It is as if their nomenclature and traditions are something to play around with, to toss around as a child tosses a toy.

Cultural appropriation concerns take place with other cultures. For instance, some Latino culture supporters have movie concerns about the "Día de los Muertos" portrayal and even about fake Mexican restaurant chains.

Asians, too, have experienced misappropriations of culture, language and prejudicial behavior and laws. Fashion with its arts and craft cultural distortions plays its role.

Sometimes with celebrations such as Halloween, we might offend someone else's culture heritage by choosing cultural regalia as costumes. Examples may include:

-Latino, LatinX celebration skirts or Day of the Dead costumes

-Asian kimonos with umbrella, ceremonial garments and other cultural regalia

-Native American, Indigenous Nations Indian costumes and feathered headdress

-East Indian wraps or sari, etc.

There are so many other prejudicial generational leftovers especially in sports names to be changed. However, it is not my intent to discuss all cultural inappropriate labels and deceptions as that would mean writing another book. It is time for these historic injustices that resulted from colonization and the dispossession of Indian lands to end.

The De-Valuing Tongue

Certain words and phrases have been habitually echoed as accepted terms and don't seem to fit into the complexity of changes for equity. It seems that these words minimize certain human beings and can be turned around and perhaps eliminated.

MINORITY - is minor, is subtraction, insignificant, a demotion. *"I am a significant African American"* rather than "a minority."

MARGINALIZED GROUPS - groups placed on the side line or on the fringe and not inclusive or part of the whole. *"I am from a LatinX neighborhood and we are whole people"* rather than "marginalized."

FIGHTING FOR PEACE - the imagery of the word "fight" connotes or suggests aggression, violence and hostility. *"I am working to obtain peace and equity"* rather than "fighting for peace." Another suggested expression is *"dissolving conflict"* or *"working with the intention to bring about peace"*

There are many other words that might be changed. What do you think about these?

MASTER CYLINDER - refers to the cylinder that is in charge of releasing hydraulic fluids to the brakes on machinery or vehicles.

SLAVE CYLINDER - refers to the cylinders at the wheels that receives the hydraulic fluids and does the work to stop the machinery or vehicles.

MASTER BEDROOM - refers to what Merriam-Webster defines as "a large or principal bedroom." However, the word "master" on its own is defined as the male head of a household and/or the owner or employer of slaves and servants.

The de-valuating tongue can come forth as **Gender Pronouns:**
Gender refers to the characteristics of women, men, girls and boys *(female and male)*. The different genders include:

Agender - a person who does not have a gender. This person may appear androgynous *(identifies as neither male or female and may physically appear as neither)*.

Cisgender -One's biological sex assigned at birth is their gender identity or perception of oneself.

Genderfluid - Depending on the day, a person may identify as either male or female, being fluid and flexible.

Genderqueer - Someone who is open about their sexual orientation. They may or may not identify as heterosexual or same-gender loving. They exhibit gender fluidity.

Intersex - A person born with chromosomes, genitalia and/or secondary sexual characteristics that contradict what is considered male or female.

Gender nonconforming - A person who either by nature or choice does not conform to gender-based expectations of society.

Transgender - This is an umbrella term for persons whose gender identity differs from the one assigned to their physical sex. Transgender persons may be straight, gay, bisexual or any other sexual orientation.

What we commonly hear and speak in relation to genders are gender pronouns.

Gender pronouns means the pronoun *(a word that can function by itself as a noun)* that a person chooses to use for themselves. The use of the term "gender binary" refers to a system in which all people are categorized as being either male or female.

Gender neutral refers to an identity for a person who does not identify as she/her or he/him.

Many people think that these gender pronouns are something new. These pronouns are being renewed or re-claimed. They have been used throughout the history of literature. As early as 1386 in Geoffrey Chaucer's *The Canterbury Tales* and in Shakespeare's *Hamlet* in 1599. The terms "they" and "them" were being used by these literary authors to describe people in the 17th Century as with Jane Austin in her 1813 novel, *Pride and Prejudice.* It was a way to specify a role being undertaken by a person.

Currently, some terms such as "ze, zir, xem" and their counterparts have not been widely used or accepted into everyday speech. I have often heard the de-valuing tongue come forth with having to use gender pronouns. "Why should I say, Kathy is so helpful because 'they' have a caring personality?"

To look on the bright side of using gender pronouns:

1) Gender neutral bathrooms in schools eliminate the pressure of trans students to conform to the gender binary.

2) A recent study found that using gender neutral pronouns reduces mental biases that favor men and increases positive attitudes towards women and the LGBTQ community. [4]

Perhaps we can learn to alleviate such statements as: *I won't say "Kathy is so helpful because they have a caring personality."*

It's important to give up complaints about speaking gender pronouns and replace these complaints with compassion.

A new pronoun story to focus on:

GENDER BINARY	Subject	Object	Pronoun	Pronunciation
	she	her	hers	as it looks
	he	him	his	as it looks
GENDER NEUTRAL	they*	them*	theirs*	as it looks
	ze	hir	hirs	zhee, here, heres
	ze	zir	zirs	zhee, zhere, zheres
	xe	xem	xyrs	zhee, zhem, zheres

*used as singular

Shopping

The de-valuing tongue can also be found during a simple shopping day. Take for instance a Black, Indigenous or Person of Color (BIPOC) individual goes to a fancy boutique where mostly well-to-do shop. The BIPOC is shopping to find that one beautiful scarf they have been saving up to purchase. At the cash register the clerk says in a high-toned, dull-sounding voice, "Is this it?" This might not be taken as a serious matter by some, yet it can sound offensive when one has happily saved for this day to match a certain outfit for a special event.

How else could a clerk speak to customers regardless of the quantity and dollars spent that would not echo as a "put-down," "less-than," "feel small" energy output?

Everyday Invisibility

Not seeing someone can be done verbally and through visual oversight. Whether conscious or unconscious, it still hurts when one feels one does not exist to another.

Take for example, a mixed-race group is having a meeting about their community center. The committee chair being a white person says, "What date did we all decide upon for our conference call?" A black committee member says, "Tuesday, the 5th." The white committee chair does not acknowledge the black person's confident and clearly heard voice. Then a white community member named Ellen says, "Tuesday the 5th" and the white chairperson says, "Thank you Ellen."

Other than verbally ignoring one's presence, there is the visually ignoring which creates a sort of non-existence. An example would be displaying and sending out wedding party photos to all in attendance. The BIPOC attendees notice there are no pictures of them.

Publicly Announcing Differences

I have witnessed so many times that it makes me feel singled out and different as though I don't belong. Looking at me one lady said in a loud voice in a crowd, "Where are you from?" Then I said, 'I live right here.' "No, where are you from originally?"

That same, "don't belong" feeling often comes up when my friends born in another country are asked: "I hear an accent. Where are you from?"

Pronounce It Like Me

I have also experienced conscious folk who cannot stand to hear me, as an African American, pronounce words when studying a foreign language.

Example: I also have observed two friends, one African American and one Caucasian, listening to a Rosetta Stone tape and learning Spanish. They then have to repeat words and sentences out loud. The excitement of learning is exhilarated. Without permission, the white friend looks at the black friend and says harshly and empathically, "No, this is how it is said…" This study arrangement thus ended as even when the black friend said, "Please let me pronounce it the way I hear it as I am absorbing the visuals as they correspond to a word, then I will focus on pronunciations. You obviously have another way that you like and are inflicting it upon me. Please do not try to control me because you are squelching my enthusiasm to learn Spanish when we do this together."

This not only happens with learning another language, it happens regularly when BIPOC individuals are corrected when they speak. It would be a great transformation if we could accept those of all races including their manner of speaking. So many precious central points of a conversation can be missed when a person blocks another's conversation to correct them the way they are comfortable in hearing it. Also, the opportunity to make a valuable human connection can be missed.

ACTION

THE NEW STORY OF HEALING & SOLUTIONS

Every solution and resource births a new story for change. If we want to see changes, we must take action.

In this part of the chapter are possible solutions and resources to help inspire actions so that we can have opportunities to live new or revived stories to:

A) Awaken to what happens when our words and expressions can harm BIPOC and vulnerable populations in both the social justice and climate justice domains

B) Find any action item as a pathway towards a climate career, social justice career or a research or volunteer project which is just waiting for you to choose from. That is, unless you have already been triggered into action.

You have read the social justice narratives such as the power and de-valuing and other narrative examples. Suggestions are enfolded that you can literally grasp or read between the lines. It is the examples themselves that bring to the surface something to think about and how narratives can be altered to create belongingness and enthusiasm. Try creating changes and influence others to do so by sharing these examples given in this chapter as possible solutions. It's a forward leap towards healing and creating racial harmony and equity for persons of color, the meek and the vulnerable with new narrative stories.

Since a lot of information has already come forth about our warming climate in previous chapters, possible solutions to changing narratives about global warming are suggested. This is also suggested in the context of listing some facts to climate deniers. Let's start off by saying this statement. That when we refuse to accept facts in order to protect us from uncomfortable truths, we are in denial.

Here are some suggestions to sift, digest and help create new narratives or stories, one at a time. It may take time for skeptics to come to one part of these suggestions. However, planting seeds for change can possibly sprout to fruition:

1) Try **using new terms** such as "changing earth patterns"

or "anthropogenic global change" instead of "climate change." Dr. George Somero comments that anthropogenic global change represent increasing ocean acidity, fallen oxygen levels, an increased hydrolic cycle, changes in wind patterns, etc.

2) **Give proven facts** from credible sources:

a) Temperatures are likely to rise 1.5 degrees Celsius within the next two decades according to the Intergovernmental Panel on Climate Change (IPCC) [5]

b) According to NASA, 2016 and 2020 have been the warmest on record. It states that "2020 temperature re cords were hit without it being an El Nino year, as in 2016 [6]

c) Climate change is detrimental to human life. Rising temperatures with growing city population of elderly have increased heat related deaths. [7]
Hyperthermia and air quality are things to focus on as temperatures rise. Air quality can affect how plants grow, how infectious diseases spread, and the harm it does to freshwater. [8]
A warming world also increases the intensity of natural disasters such as wildfires, hurricanes and our winters. According to World Meteorological Organization (WMO), the burn area and the intensity of fire has increased as fires have ripped through California. [9]
There is the frequency of increasing hurricanes ranking in the top categories of 4 and 5. This increase has been happening over the last 30 years. [10]
Winter sports are affected by the loss of ice and snowy winters. [11]

d) Almost half of all amphibians are at risk of extinction due to climate change. [12]

e) There is a need for climate literacy to be taught in K-12 curricula to help teach about what's happening on earth we all share and what solutions can be acted on. 13

Other than narrating reputable facts of environmental change and implementing new terminology to deliver to skeptics, lets **listen**. We can help change deniers by listening to their concerns. Listen for something that you can connect with and have another conversation at another time and gradually open doors without arguments or debates. Without awareness that there is an issue that we can all help curb, there cannot be solutions. Here are some suggested questions when having a tea or at a gathering or unexpectedly having the topic of global warming coming up: [Listen without commenting as you ask a question that could be any of the following or another that fits].

- You seem very concerned about this topic. What worries you about global warming or climate change?

- What do you think about having electric vehicles, hybrid vehicles, or even LED lights?

- We have such a beautiful environment. How do you think we can preserve it?

Let's not forget that any of these actions will also help:
- the winged ones
- the crawlers
- the swimmers
- the hoofed and claw-toed ones
- the microscopic ones

Create a list of Action Items you might change or list any interesting points.

Justice

Chapter 4

Measuring CO_2 And Racial Discrimination

There are many layers of focus and action we have to continue to consider that mirror social justice and climate justice. The area of measurement is also a key factor. Quantifying weight, area, volume, length and temperature in a scientific sense, is what usually comes to mind when we speak of measurement.

When we look at the benefit or value of measuring anything, whether CO_2 (carbon dioxide) in the atmosphere or racial discrimination in our midst, it is all important.

Measurement is like taking the proper dose of medicine. To treat the situation, the benefit comes from knowing if there is a little amount of action needed to remedy a situation. Figuring out if much more action or doses of medicine are required is also beneficial. In other words, how much CO_2 do we have to reduce in the air as well as reduce racial discrimination in the atmosphere.

By taking a peek into the climate area of measuring carbon

dioxide in the air or atmosphere, we interpret this element as something that gives us room to expand our lungs and breathe. The problem, as you well know, is the air is being polluted with too many greenhouse gases often referred to as the greenhouse effect. To explain it in simple terms according to National Aeronautics and Space Administration (NASA), one can think of the greenhouse effect as a greenhouse for raising plants. Imagine that planet Earth were sitting inside a greenhouse instead of plants being there. The glass or plastic of the greenhouse traps in the Sun's heat to keep the plants warm and grow. Well, if the Earth were the plants, it would be getting really hot, as Earth's atmosphere traps the Sun's heat. 1

Luckily, some of the gases go off into the atmosphere surrounding Earth, radiating from Earth towards space, absorbed by the lands and ocean heating the Earth. Yet, other greenhouse gases stay in the Earth's atmosphere. It is that stay, which is the problem.

This build-up of gases in our Earth's atmosphere as explained previously in the IPCC 's Summary for Policymakers report (Chapter 1): *Human activities are estimated to have caused approximately 1.0 degrees Celsius of global warming above preindustrial levels with a likely range of .8-1.2 degrees Celsius.*

According to the Environmental Protection Agency (EPA), humans are emitting greenhouse gases into the earth's atmosphere, especially carbon dioxide, that contribute to the greenhouse effect. These heat trapping gases include: 2

Water Vapor (H_2O)
Carbon Dioxide (CO_2)
Methane (CH_4)
Nitrous Oxide (N_2O)
Ozone (O_3)
Chlorofluorocarbons (CFC)

This takes us to the inquiry of measuring CO_2 levels in the atmosphere. If we are going to talk about the need to decrease greenhouse gases, we have to know how much is in the atmosphere by way of measurement on any given day.

The late scientist, Dr. Charles David Keeling, who was on the faculty of Scripps Institution of Oceanography, UC San Diego, presented the first evidence that carbon dioxide produced by automobiles and factories was negatively affecting Earth's climate.

Dr. Keeling created a way to measure CO_2 in the atmosphere called the Keeling Curve. He first began collecting carbon dioxide samples in 1958. Dr. Keeling did his first samplings in Big Sur at Pfeiffer Big Sur State Park, where he loved to camp in clean air, before going onto large scale samplings atop Hawaii's Mauna Loa Laboratory.

To explain further, the Keeling Curve is a graph of the accumulation of carbon dioxide in the Earth's atmosphere. It is based on continuous measurements taken at the Mauna Loa Observatory from 1958 to the present day. Dr. Keeling started the monitoring program and supervised it until his death in 2005. 3

Dr. Keeling measuring CO_2.

At the Mauna Loa Observatory, one can track the increments of CO_2 in the atmosphere. There, Dr. Keeling and staff collected very clean bottles of air without industrial interference.

Dr. Keeling's undulating, yet continuous, rising lines on his curve, shows strong seasonal variations in CO_2 levels reaching its peak in late northern hemisphere winter and reducing in the spring and early summer each year. This is mainly due to the increased plant growth in the fertile lands of the northern hemisphere.

Keeling's Curve actually evidenced that the earth is really

getting hotter. This would affect how the ecosystems (biological community of organisms) and climates react and show up with extreme changes in our global weather patterns.

In order to see the current day's measurement of CO_2 measurements, please search Keeling Curve Current Day Measurement. 4

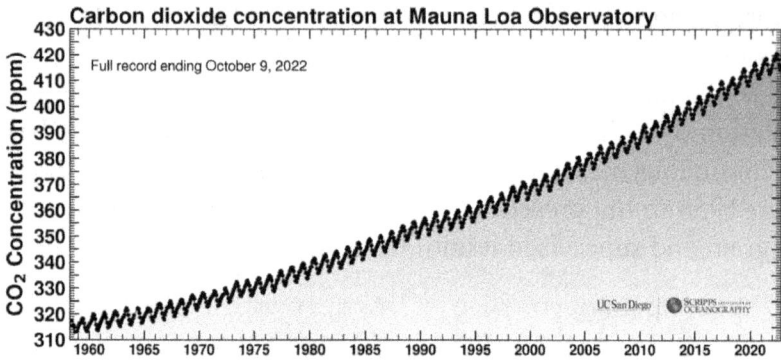

Carbon dioxide concentration at Mauna Loa Observatory

Full record ending October 9, 2022

**Keeling Curve Graph-Courtesy UC San Diego,
Scripps Institution of Oceanography**

There is a video to further examine to learn how carbon dioxide is measured. Search for this using the video title – How Scientists Measure Carbon Dioxide in the Air Scripps. 5

Bear in mind what Dr. Keeling did was revolutionary; to measure the carbon dioxide level in our atmosphere. It drew a lot of conflict as it became a worldwide political debate over the reality of global warming with correct measurements. He not only measured it; he is also in the category of a "bell ringer" by giving the world early warnings about the dangers of human induced climate change.

We live and breathe CO_2 Do we live and breathe racial discrimination? Can it be measured like CO_2? If only there was a clear-cut way to actually calculate and quantify racial discrimination, it would be quite useful to start doing much more than is already done to eradicate it. Now that we have measured CO_2 in the environment, we have started the awareness of climate justice as pointed out in Chapter 1. Let's explore a little bit about measuring discriminations as concerns one's race.

In a study by Bar-Anan and Vianello in 2018 there was an effort to measure racism. According to Ulrich Schimmack, who wrote an abstract on this Implicit Association Test (IAT), the data critique as a measure of implicit bias assessed political orientation and racism with multiple measures. Participants reported whether they voted Republican or Democrat. Only White participants who reported voting were included in the analysis. The results show the strong relationship between voting and the Republican factor shows the political orientation can be measured well with a direct question. It also implies that only a third of the variance or difference in the actual rating reflects racism. The rest is variance error. In other words, there is no perfect way to measure racism. 6

When researching racial discrimination, the thing called Artificial Intelligence (AI) is often biased against a person or group. Al is what is being referred to as the programming language that is used for manipulating data that can be retrieved, inserted or modified. In many instances' computers are used to modify human behavior. This type of Artificial Intelligence is used a lot with existing human bias transferred to the programming. It started quite some time ago. Let's look at a few examples of AI being racially prejudice or biased.

According to World Economic Forum, a few examples were given. 7

Back in 2014 Amazon wanted to review resumes of job applicants using Al. By 2015, the company realized its new system was not rating candidates for software developer jobs and other technical posts in a gender-neutral way. That is because Amazon's computer models were trained to vet applicants by observing patterns in resumes submitted to the company over a 10-year period. Most of the software developers were men. This mirrored what was happening across the technological industry. Of course, when it was discovered that the system discriminated against women for technical roles, the software was exited. 8

Another example of Al being biased is through its use in 2019 by San Francisco lawmakers. These lawmakers voted against the use of facial recognition believing they were prone to errors when used on people with dark skin or women.

According to a Harvard University blog on science policy and social justice, when AI was used in law enforcement, there were racial biases that caused harm to BIPOC communities due to its inaccuracies. In some cases, there have been a mis-identification of suspects leading to the incarceration of innocent Black Americans. Additionally, face recognition can potentially target other populations, such as undocumented immigrants by ICE, or Muslim immigrants by the police.

As stated by the Algorithmic Justice League, "face surveillance threatens rights including privacy, freedom of expression, freedom of association and due process." 9

Biased AI is found in mortgage loans involving what is considered by many as a minority group. This is because minority groups have less data documented in their credit histories. The flaws of artificial intelligence on BIPOC in healthcare shows a system that favors white patients over black patients in predicting patients that need extra medical care. 10

So where does measuring discrimination and racism measure up to being equitable for people of color? One might conclude that it is an insignificant issue to those who develop algorithms used to create artificial intelligence. After all, it took a lot of research and motivation to come up with artificial methods of measuring to be objective and yet racism and discrimination are still not dealt with fairly by mankind nor its machines. Changes have to come to make this work for all people.

We know racial discrimination exists so let's get it right and lets all get on-board.

Even after, should it ever happen, that racial discrimination is quantified as an extreme priority to dissolve on this planet, we have a lot to do. Are we all willing to heal our past traumas, shame, hurt and guilt and become open hearted and action-oriented for the rest of our lives to dissolve racial discriminations?

ACTION

THE NEW STORY OF HEALING & SOLUTIONS

Every solution and resource births a new story for change. If we want to see changes, we must take action.

In this part of the chapter are possible solutions and resources to help inspire actions so we can have opportunities to live a new or revived story to:

A) To learn about quantifying or identifying if very little or a lot of action is needed in order to remedy or reduce CO_2 and racial discrimination in the environment

B) To find any action item as a pathway towards a climate career, social justice career, research or volunteer project that is just waiting for you to choose from. That is, unless you have already been triggered into action.

List of possible action items to pursue to reduce racial discrimination in the environment:

- Offer support to the victim. Listen carefully and respect confidentiality.

- Speak up or seek help when you experience discrimintion. Recognize that some situations are best addressed publicly and others privately.

- Be a witness to any form of racism by filming policemen and other officials, misusing their powers and mistreating people from minority groups

- Catch yourself judging other people

- Become involved and work with others on an action item(s)

- Encourage work and study environments to be places where diversity is valued.

- Discuss issues of inclusion and diversity with children, youth and adults.

- Observe yourself judging others

- Gather your employees, staff or organization to register for a Bridge Building to Equity Workshop. Contact info@lavernemcleod.com or call (831) 595-9692 or research other Inclusion workshops.

Here are suggestions to help in reducing greenhouse gases in the atmosphere:

- If you are thinking about doing research with a Hawaiian Observatory:
  Search: **Observatory Jobs** in Hawaii [11]

- To help in continuing with Dr. Charles David Keeling's work, you can research the various Keeling Prize participant projects and winners of this prize started by Dr. Keeling's son, Dr. Ralph Keeling. It is an annual prize conceived by philanthropists through the Scripps Institution of Oceanography, University of California San Diego with the Keeling family's permission. The mission prize is greenhouse gas mitigation (lessening) and carbon intake to reduce heat trapping gas in the atmosphere. Research if you could help one of the prize winners who have already started on the path of greenhouse mitigation or how to compete for this prize.
  Search **Global Warming Mitigation Project** [12]

Let's not forget that any of these actions will also help:
- the winged ones
- the crawlers
- the swimmers
- the hoofed and claw-toed ones
- the microscopic ones

What is the measurement of CO_2 where you live?

Electric Charge

Chapter 5

Transportation

As we move around in our vehicles, aircraft, ships, boats, rails, construction and agricultural equipment, we often take for granted that everyone can do the same. We note here that these methods of transportation add to greenhouse gases in the atmosphere. On the other hand, not everyone has access to these modes of transportation. They rely upon public transportation that reduces greenhouse gases yet increases in racial inequity. This is the intersection mirroring climate justice and social justice.

Most of the information about the transportation industry doesn't matter because in many BIPOC and poor communities, folks cannot afford them. Can you see an elderly black woman needing to carry her groceries home on a scooter and there is not a way that she can ride-share either because she doesn't have a ride? Ride sharing apps are thus, out of the question.

The outbreak of COVID-19 even revealed more dents in the transportation system. Folks had to potentially expose themselves to the virus or a possible carrier. This meant less folks used the public transportation system or they risked their lives to feed their families as low-income, frontline workers and others could not consider the "stay-at-home" orders.

To compound the matter, living in neighborhoods where violence is prevalent can create difficulties when waiting for public transportation or buses. A good or at least an average transportation system in any metropolitan area, large or small, possesses the power to create a path for everyone to access transportation to and from home, work, obtaining food, education, and healthcare. The problem is there is not a "seat at the table" to represent this population. Even if one was interested to go to a meeting about transportation in their neighborhood, they might lose their low-income job that they count on to feed their family and keep a roof over their heads. More than often, money is prioritized to go into roads and highway construction that serve those privileged to have a car and other transportation vehicles rather than urban transportation systems.

Pew Research Center indicated that nationwide, 23% of transit riders are Black, 15% are Latino and 7% are white. Approximately 15% of all transit riders make less than $30,000 annually. A specific example is what occurs in Los Angeles, California according to PEW.

It has the second-highest public transit use in the U.S. and the racial and wealth gap is significant. The Los Angeles County Metropolitan Transportation Authority's (Metro) 2019 survey of its riders found that 66% of bus riders are Latino, 15% are Black, 8% are white and 7% are Asian. About 57% of riders reported that they were below the poverty line. The median household annual income for bus riders was $17,975. Generally, more Latinos and low-income individuals ride the bus than take the rail. The bus is the second-most used form of transportation in Los Angeles—second only to cars. 1

To further support previous statements is the Brookings Institution research group's Metropolitan Policy Program. Brookings report that one possible reason that Blacks, Hispanics and immigrants might be bigger users of public transit is because they are more likely, than Americans overall, to live in large metropolitan areas where there tend to be more public transit options. They are also less likely to have access to an automobile than other groups and are more likely to use public transit for commuting to work. Blacks and Hispanics also tend to live farther away from their jobs, which could make walking or biking to work less common. 2.

According to an abstract from the Journal of the American Planning Association, "Urban transit systems in most American cities...have become a genuine civil rights issue-and a valid one-because the layout of rapid-transit systems determines the accessibility of jobs to the Black community. A good example of this problem is the city of Atlanta, where the rapid-transit system has been laid out for the convenience of the white upper-middle-class suburbanites who commute to their jobs downtown. The system has virtually no consideration for connecting the poor people with their jobs." 3 In addition, Atlanta, Georgia, Portland, Oregon and in many if not most cities in the United States have transportation systems that do not work well for poor folks.

When we look at European and other international places, Bloomberg City Lab states "It's different over there. Europe is far more densely built, and its older cities—settled centuries before the automotive age—will always be innately transit-friendlier. In Asian cities, meanwhile, explosive urban growth has been accompanied (and accelerated) by massive government investments in urban rail networks. But the U.S. boomed in the 20th-century's automobile age, and the private car is still king." 4

By the same token, let's pivot our reflection to cars, trucks, aircraft, ships/boats, rails, construction and agricultural equipment. A United States' Environmental Protection Agency (EPA) study on greenhouse gas emissions notes the following information:

"The majority of greenhouse gas emissions from transportation are carbon dioxide (CO_2) emissions resulting from the combustion of petroleum-based products, like gasoline, in internal combustion engines. The largest sources of transportation-*related greenhouse gas emissions include passenger cars, medium- and heavy-duty trucks, and light-duty trucks, including sport utility vehicles, pickup trucks, and minivans. These sources account for*

over half of the emissions from the transportation sector. The remaining greenhouse gas emissions from the transportation sector come from other modes of transportation, including commercial aircraft, ships, boats, and trains, as well as pipelines and lubricants. The EPA also states that the main human activity that emits CO_2 is the combustion of fossil fuels (coal, natural gas, and oil) for energy and transportation, although certain industrial processes and land-use changes also emit CO_2.

Relatively small amounts of methane (CH_4) and nitrous oxide (N_2O) are emitted during fuel combustion. In addition, a small amount of hydrofluorocarbon (HFC) emissions are included in the Transportation sector. These emissions result from the use of mobile air conditioners and refrigerated transport." [5]

Speaking of ships: Royal Caribbean Cruises Ltd. recently announced a special five-year global partnership with the World Wildlife Fund (WWF). This is an initiative designed to help ensure the long-term health of the oceans. The cruise line, working with WWF, is setting measurable and achievable sustainability targets that will reduce Royal Caribbean's environmental footprint. The article goes on to say that a few cruise ships are offering passenger tours to their waste management decks." To further the effort for environmental stewardship, the focus is energy and air emissions; water and wastewater; waste and chemical management; conservation, destinations and education; community involvement; safety and security; medical operations; public health; and human resources - a move towards transparency. [6]

There is a big focus on electric vehicles. They are very responsive, have very good torque are often more digitally connected than conventional vehicles. Many electrical vehicle charging stations provide the option to control charging from a smartphone app. One may never need to go to a gas station again. [7]

Other than individual benefits, electric vehicles can help the United States have a greater diversity of fuel choices available for transportation. The U.S. used nearly nine billion barrels of petroleum last year, two-thirds of which went towards transportation. [8]

Concerning high speed rail, the Association of American Railroads, moving freight by rail instead of trucks, lowers greenhouse gas emissions by up to 75 percent, on average. Also, railroads are three to four times more fuel efficient than trucks. A single freight train can replace several hundred trucks, freeing up space on the highway for other motorists. Shifting freight from trucks to rail also reduces highway wear and tear and the pressure to build costly new highways. 9

ACTION

THE NEW STORY OF HEALING & SOLUTIONS

Every solution and resource births a new story for change. If we want to see changes, we must take action.

In this part of the chapter are possible solutions and resources to help inspire actions so we can have opportunities to live a new or revived story to:

A) Address the transportation industry to see how to reduce racism and greenhouse gases as one issue to be resolved.

B) Find any action item as a pathway towards a climate career, social justice career, research or volunteer project that is just waiting for you to choose from. That is, unless you have already been triggered into action.

List of possible actions to pursue to reduce racism in public transportation:

• Help pursue what many states are doing to make it possible for undocumented immigrants to own or drive a car. According to NCSL (National Conference of State Legislatures), sixteen states in 2021 have made this possible.

These states—California, Colorado, Connecticut, Delaware, Hawaii, Illinois, Maryland, Nevada, New Jersey, New Mexico, New York, Oregon, Utah, Vermont, Virginia and Washington—issue a license if an applicant provides certain documentation, such as a foreign birth certificate, foreign passport, or consular card and evidence of current residency in the state. 10

Search: States Offering Driver's License to Immigrants

- Ask for, petition, publicize, obtain allies to address community public transportation concerns. Get a "seat at the table" for transportation authority meetings. Be sure that the public transportation system be laid out to provide an opportunity for poor people to get meaningful employment in order to move into the mainstream of life.

- Become an advocate to remind "lawmakers to be mindful to not create a mass transit system that only accommodate the comforts of the *affluent*, who are most often white." If U.S. lawmakers and voters can come to terms with systemic racism's role in stifling mass public transporttion, one can begin to see economic and social benefits for everyone—not just for people who are transit-dependent. 11

- Think about what the late Reverend Dr. Martin Luther King, Jr. had to say about public transportation that relates to people of color. "There is only one possible explanation for this situation, and that is the racist blindness of city planners." Again, be an advocate or an ally, go to city planning meetings, speak up to city planners and invite others to join you.

List of possible action to pursue to reduce greenhouse gases in transportation:

- Explore-Careers in Electric Vehicles on the U.S. Bureau of Labor Statistics official website of the U.S. government: 12

1. Explore- <u>Wright Electric</u> whose goal is to reduce emissions in aviation. This company is a Hybrid Electric Airplane manufacturer with swappable battery packs allowing for hybrid electric and fuel travel. Thus far, they are building electric planes that lower fuel costs, noise, emissions and runway takeoff time. 13 Contact: Jeffrey Engler- 917 608 9785 contact@weflywright.com

- Explore- Pipistrel is a producer of small scale/small passenger airplanes. They build Alpha Electro airplanes that are fully electric and designed for two people training flights. Contact: (213) 984-1237 for USA; Contact: 14 5522 0583 for AU; info@pipistrel-usa.com https://www.pipistrel-usa.com/

- Explore- more information about companies on this public website devoted to reduce greenhouse gases in Electric Vehicles, Shipping Transportation and transportation accessories, etc. includes over 100 listings to research: Search : EV100-Climate Group Members 15

- On the Green Vehicle Guide you can search for green vehicles and see information on light duty vehicles, including emerging vehicle technology and alternative fuels. The site also addresses transportation's role in climate change. Search: United States EPA-Green Vehicle Guide 16

Suggested ways of reducing emissions from transportation:

- Using public buses that are fueled by compressed natural gas or electricity rather than gasoline or diesel.

- Using electric or hybrid automobiles, provided that the energy is generated from lower-carbon or non-fossil fuels.

- Using renewable fuels such as low-carbon biofuels.

- Developing advanced vehicle technologies such as hybrid vehicles and electric vehicles, that can store energy from braking and use it for power later.

- Reducing the weight of materials used to build vehicles.

- Reducing the aerodynamic resistance of vehicles through better shape design.

- Reducing the average taxi time for aircraft.

- Driving sensibly (avoiding rapid acceleration and braking, observing the speed limit).

- Reducing engine-idling.

- Improved voyage planning for ships, such as through improved weather routing, to increase fuel efficiency.

- Building public transportation, sidewalks, and bike paths to increase lower-emission transportation choices.

- Zoning for mixed use areas, so that residences, schools, stores, and businesses are close together, reducing the need for driving. Search for EPA Sources of Greenhouse Gas Emissions 17

Let's not forget that any of these actions will also help:
- the winged ones
- the crawlers
- the swimmers
- the hoofed and claw-toed ones
- the microscopic one

Jail

Chapter 6

Part I
Disposable World

    This is an extensive two-part chapter due to the nature of how human beings, especially BIPOC are treated in the United States of America and other parts of the world. Jane Elliott, a white diversity educator gives a talk on "Being Black" to a white audience, made a statement: "White People Are Not Ignorant about Racism/White Supremacy." To support her statement, she made a request to her audience:

    "I want every white person in this room who would be happy to be treated as this society in general treats our Black citizens to please stand." (pause). No one stood. Elliott continued to say, "You didn't understand the directions- If you want to be treated like Black folk in this society, stand." (pause) "Nobody is standing here. That says very plainly that you know what's happening. You know you didn't want it for you. I want to know why you are so willing to allow it to happen to others?" 1

Just pause and take a deep breath. This is a story that has to be changed. The next story has to be changed as well is about breathing. This phrase that most of the world is familiar with is "I can't breathe," George Floyd's last dying words. Floyd was publicly executed by a police officer as he held his knee on Floyd's neck, helpless, handcuffed and on the ground. Rings true to lynching with hands tied on backside as a rope that rapidly squeezes life force from innocent black men and women. This beat historically continues. It is like the invisible speaking. Those people are considered throw-aways, discarded and are thus disposable.

Consequently, other methods of people disposal are happening simultaneously with business and household garbage toxins and pollutants. Getting rid of unwanted chemicals and trash affects both people and planet, one of the main focal points of this chapter.

A quote from an article "Racism is Killing the Planet," by Hop Hopkins, tells the current story. Hopkins is an African American Environmentalist and Director of Organizational Transformation and an Interim Executive Steering Committee member at the Sierra Club.

"In order to treat places and resources as disposable, the people who live there have to get treated like rubbish too. Sacrifice zones imply sacrificed people. Just think of Cancer Alley in Louisiana. Most of the towns there are majority Black, and nowadays they call it Death Alley, because so many black folks have died from the poison that drives our extractive economy. Or think of the situation in the Navajo Nation, where uranium mines poisoned the wells and the groundwater and coal plants for decades poisoned the air. Or consider the South Side of Chicago, where I used to live, which for years was a dumping ground of petroleum coke (a fossil fuel by-product) where residents are still struggling against pollution-related diseases. I've lived in a lot of places, and just about every place I've ever lived has been targeted by big polluters as a dumping ground." 2

Harriet Washington, a lecturer in bioethics at Columbia University, along with many other titles, writes about environmental racism in regards to the IQ of individuals who have been exposed to toxins in the environment, as well as, in their drinking water. She indicates that people of color are at a disproportionate risk with the scourge of air pollution. They lose breath and brain.

This is also happening in other countries. The World Health Organization (WHO) found that more than four out of every five urbanites on the planet, and mostly in the developing world, live in neighborhoods where air quality falls below minimal health standards. Karachi (southern coastline of Pakistan), Lagos (Nigeria), and Beijing (northern China) cities are cities notorious for their visible smog. Their citizens breathe poisonous chemicals and brain-draining particles. 3

The developing brain of children are injured with air pollutants according to Harriet Washington who authored *A Mind is a Terrible Thing to Waste*. She explains that children have a greater lung surface area relative to their body size. This gives them greater relative exposure to noxious gases and suspended particles than adults. This increases risk for fetuses and infants.

In California, the coal industry is taking its toll on Black and Latinx communities. You know what happens when you wake up on a certain morning in an area where snow is expected? Do you rejoice at seeing the first signs of snow on the ground? Imagine waking up to something else on your lawn -a black dust. Look out your window sill or at your automobile. You see the same thing everywhere. This happens at least twice a week in a historically Black neighborhood. It's Parchester Village in Richmond, California. This also occurs in the neighboring Santa Fe and Nystrom Village where Latinx predominately live. One hundred box cars or train loads of coal mined from the mountains of Utah get unloaded at a privately-owned coal shipping port, the Levin-Richmond Terminal. From there, it is shipped out to various places and countries.

You might wonder why Black and Latinx don't move from those communities. We can guess that the main factor is economics, trapped in a poor box, almost impossible to get out of. This entrapment came in the midst of World War II. Black people migrated to the area from the American South to work in the shipyards. Housing discrimination forced them into the city's most polluted industrial neighborhoods where they still live today. Latinx have a longer history of being in Richmond, CA, beginning to live there in the early 1800's. 4

According to the Centers for Disease Control and Prevention (CDC), the State of California has the highest current number of persons with asthma. 5

Here are some comparisons:

Alabama 360,965
Alaska 54,239
Arizona 544,394
Arkansas 215,095
California 2,405,797

Richmond, California is the pollution hotspot in the County of Contra Costa. It has the highest emergency department visits for asthma-related symptoms in California. Thus, it is a burden on this county where pollution-creating refineries and terminals exist. It is where the population is 18.2% Black and 44.1% Latinx according to the US Census Bureau report for the city of Richmond, CA. 6

How can one say, "I care about saving the planet," when a certain percent of the people living on it are considered waste?

There often is a lack of trees in industrial and poor neighborhoods. There, too, are vacant lots resulting in large blocks of pavement which absorb a great amount of heat during very hot days. Here, urban planners do not showcase how green space captures carbon. When there's a lot of rain in poor neighborhoods, where are the green patches to mitigate flooding? How come there are electric transformers and high voltage electric towers routed in these neighborhoods?

Efforts to hear Black voices via organizations such as Future Human and No Coal Coalition have opened doors and have been praised as advocates to stop the polluting coal entities that are ruining the health of people. However, these doors are closed by opponents filing lawsuits and being derailed by higher courts consistently favoring coal companies and white affluent coalitions of the "not-in-my-backyard" (NIMBY) movement. 7

Randolph, Arizona is an unincorporated and economically depressed community located 60 miles southeast of Phoenix. Polluting industries there are rising. The community was founded a century ago by Black migrant farmworkers. These migrant farmers used to chop and pick cotton. However, that industry does not exist anymore. Folks are starting to move out of that area. Yet, there are those who call it home and are staying in their tattered, poverty-ridden houses.

A major project that includes several gas-burning plants are endorsed by Randolph's nearby city of Coolidge. This city stands to reap millions in tax revenue. This project also meets the power demand of the region of Phoenix. Those black farmers who remain, as they have for generations, do not have the resources to contest the situation nor relocate. Breathing pollutants damaging to the climate and their bodies, they stay and become disposed persons. 8

Back in 2012 the National Association for the Advancement of Colored People (NAACP)'s report titled *Coal Blooded* states that "of 378 coal plants studied, the 12 identified as the worst polluters were located in or near cities with large Black populations. This included Detroit, Chicago, Cleveland, and Milwaukee. In these cities, 76% of the population living near the plants were people of color. These plants emitted copious amounts of sulfur dioxide and nitrous oxides. Inhaling these compounds is linked to asthma, bronchitis, nausea and vomiting, respiratory infections, decreased fertility, and harm to fetuses." 9

To look at this same study in 2021, all but two of the 12 worst-polluting coal plants in the United States either have shut down, plan to shut down, or have been converted to natural gas. However, natural gas is a major concern because it is mainly methane, a strong greenhouse gas. 10 Methane, too, is a pollutant that leaks in abandoned coal plants. According to the EPA, abandoned mine methane (AMM) is found in diffused vents, ventilation pipes, boreholes, or fissures in the ground. 11

Stored coal ash is prevalent on the power plant facility grounds long after they are shut down. It seeps through the ground into waterways humans rely upon. This underground water seepage plays out over and over again. This is evidenced in the Black community of Mattaponi Creek in Maryland. Running through the community is 217 acres of coal ash landfill from the Brandywine Ash Management Facility. Flowing from Mattaponi Creek into the Patuxent River is water to thousands of people that is full of contaminants.

There's still the wait and see factor at work. In early 2021, Maryland's state legislature was working on a bipartisan bill to phase out the state's coal-fired power plants. At the same time the bill is aimed to curb economic impacts and job losses due to closures. This is according to Maryland Matters, a nonpartisan energy and environment publication. However, the bill was withdrawn in March. Labor unions wanted to continue to collaborate with environmental activists on a transition plan. It is wait to see where this bill goes. [12]

"Environmental racism," for the polluting of Black and Latinx communities in the city of Oakland, California is an ongoing term applicable to these areas. Plans were announced in 2015 for building a huge shipping facility called Oakland Bulk and Oversized Terminal in the West Oakland area. This would give coal companies in landlocked states the ability to ship overseas. However, it would only worsen the air quality for the nearby Black and Latinx communities. Isn't it enough that these communities are already bordered by three freeways surmounting to the worst air quality in California according to a California Government environmental screening report? [13]

Hop Hopkins, referred to earlier in this chapter, continues to say in his splendid and yet painfully truth-telling words: *I wish I had all the answers, but I don't. The answer is for all of us to figure it out together. All I know is that if climate change and environmental injustice are the result of a society that values some lives and not others, then none of us are safe from pollution until all of us are safe from pollution. Dirty air doesn't stop at the county line, and carbon pollution doesn't respect national borders. As long as we keep letting the polluters sacrifice Black and Brown communi-*

ties, we can't protect our shared global climate."

So many people are exposed to harmful pollutants. According to National Institutes of Health's (NIH) National Library of Medicine, there are immediate and long-term health consequences of exposure to agricultural chemicals among migrant and seasonal farm workers that happens accidentally.

Farming has allowed mankind to harness various environmental resources which feed us. This can also be problematic as a NIH abstract points out. "It is the disruption or interference of the natural ecosystem by farming and the large population concentration leading to a variety of health threats to humans." 14 Many chemicals are used to prevent crop failure. These chemicals include pesticides, insecticides, herbicides, fungicides and others. They come in the form of gas, liquid, granular or dust applied through spraying and crop dusting.

The labor force of migrant and seasonal farm workers has little power to control their exposure to chemicals. A large percentage of fruits and vegetables produced in the United States are hand-harvested or hand- cultivated. Workers arrive predominately from Mexico with solid agricultural skills firmly grounded in practical experience and with a working knowledge of agriculture. They are young, averaging about 31 years of age. Some arrive as single men. It is estimated that between 1 and 3 million migrant farm workers arrive in the USA every year according to a State Diversity Specialist of Cornell University Cooperative Extension. He also indicates that more than half of all farm workers – 52 of every 100 – are unauthorized workers with no legal status in the United States. These workers have a strong work ethic, agricultural know-how and commitment to produce and harvest non-blemished perfect fruits and vegetables.

Migrant workers also come from countries such as Jamaica, Haiti, Guatemala, Honduras, Puerto Rico, the Dominican Republic, and other states in the United States. They leave what is familiar and comfortable or dangerous to seek their hopes and dreams to make enough money to support their families back home; feed themselves; purchase land and a home. Many immigrants ultimately return to their homeland, not realizing the health risks involved. Another factor for migrants wanting to return to their homeland is the low wages and high rents.

The health effects, depending on the type of farm chemical used, begins to show up through contact with the skin, the respiratory track, the eyes and the gastrointestinal system. This results in neurological issues as anxiety, memory defects, mood changes, visual impairments, reproductive problems (sterility, spontaneous abortion, birth defects), increased risk of cancer and even death.

Beyond health risks is the fact that the lifestyles these seasonal and migrant farm workers live is that of isolation and not being around their families. This is a big disparity that society has imposed in order to get these workers, who sometimes don't fulfill their dreams and instead return home to poverty. These folks are like invisible and uncounted human beings that are being used without personal connection or inclusion with language barriers, culture starvation and empty-hearts of deserting their families and feeling thrown away. 15

Unfortunately, there are other ways that people of color, vulnerable and poor people are systematically casted out on Planet Earth. Not all ways are covered in this book. To mention some of the ways, let's consider the prison system and the homeless.

Starting with the homeless, a visible story, similar to field workers that we just quickly gazed at while driving by a field. Being homeless, without warmth and proper shelter, a basic human need is deprived. You probably see it all the time, make a critical judgment call and keep moving on. This is what I have done in many instances until one day, I changed my mental action. I believe that one has to have the mental action empathically conceived in the mind before true physical action can take place to help. Here is my story:

'My golf group having held an annual Christmas party at the Monterey Yacht Club had a cheerful gathering. *[FYI- my golf group membership raises money for the local Food Bank and purchases holiday gifts to elder homeless women and other homeless families.]*

Afterwards, I cheerfully walked outside our golf party. I saw sitting on the ground against a nearby building, a very elderly homeless man. His bags of belongings lay soaking in the sun. I smiled and said hello. He seemed to be in another world. I thought about the unfair distribution of wealth in affluent Monterey County. Does he get meals? What happens when it rains or the temperature drops? This broke my heart even though I have witnessed this many time before. Since I had just published my company's December newsletter titled, "Don't Forget the Elders During the Holidays," my mind was wide open after seeing this particular homeless elder. My newsletter purpose is to present and inspire others to be in action for change. This led me to do further research which add to the disposable aspect of this chapter. 16 Interest was stirred by some subscribers. One subscriber, Monica Eastway, a local eco-gerontologist, made a great case of an alternative to institutional elder care. 17

What I couldn't get out of my head is what my friend, Anjani Belisle, another subscriber, said about the words *homeless* and *elder.* "They should not be in the same sentence." I considered it could be an acronym, NIIS, not in the same sentence.

Shortly after our initial conversation, I received an email from Anjani who wrote to eight of her friends who live in Hawaii:

"Dear friends,
Please let me introduce you to a longtime friend from Big Sur, CA, LaVerne. She is love in action.
I know just yesterday we looked at this issue with someone we know; whose bags were packed and she had nowhere to go!
It was heart breaking to go into Kahalui after the storm and see people on the side of the road under garbage bags.
This issue is not "out there". Let's get together soon and see what we come up with and how we can take action.
It doesn't have to be.... taking on the homeless problem here on Maui in a big way...we can take small steps and let it grow.
love you
Aloha
Anjani"

I then called for a Zoom discussion meeting *Free Zoom to Dis-*

cuss Homeless Elders to get up to speed on what is really happening in Monterey County that might apply to other places as well. Outside one of my golf group's recipient agencies we filled one room full of holiday gifts for homeless elders and families, there I met someone who was filled with knowledge about the homeless in Monterey County. She is a "boots on the ground" angel. At a few homeless encampments who trust and know her. She provides them with holiday meals and clothing, procured by a local restaurant kitchen donated especially for them. This angel attended our Zoom. There were other interested attendees as well. The following questions were the focus:

- What are the great things happening in Monterey County to help homeless elders?

- Can anything else be done?

- Anything missing?

- How can we bridge any gaps?

It was in the missing gaps where the attendees dug deeper-from COVID and personal safety to violence were among the concerns. The outcome of this Zoom, as far as action solutions are concerned, can be reviewed in the Action part of this chapter.'

Not stopping there, I contacted a local friend to help find homeless persons in Big Sur. CA. We only identified three persons. My friend and I talked about what we might do. I was busy writing this book. There are resources available in Monterey County that actually helps homeless individuals in the action section of this chapter. I rest on those resources for now. I did not have time to create what I will someday get back to called NISS, (not in the same sentence) inspired by my friend.

Another way people of color, vulnerable and poor people are systematically disposed of or thrown away on Planet Earth is by locking them up in jails and prisons. When trying to understand this concept of "disposal," we uncover a story of locking up folks for-profit that contribute to making big money for private prisons and its shareholders. Black and Latinx men are often prevented from achieving their potential. The fear-of-retaliation for past tortures and terror of enslavement is unconsciously enforced. Could it be said that the "war on drugs" was a rubber stamp of approval guaranteeing lock-up of mostly black men? There were unusually harsh sentencings targeting prisoners from black and poor communities. 18

Consequently, the prison system is a way to punish and not deal with what is really going on behaviorally to a human being who ends up incarcerated. One may wonder how this system works. Since this chapter does not completely cover the prison system, I touch upon the subject because it the potential of valuable human lives that cannot always manifest that needs to be considered and hopefully not be deleted from our thoughts.

The term "school to prison pipeline" is like an echo of the "Black Codes" originally designed to keep the black man down and oppressed. The Black Codes were restrictive laws designed to limit the freedom of African Americans who coughed at the wrong time or didn't move from the sidewalk for a white person, thus being arrested for minor crimes. These types of crimes are followed through in today's society with misdemeanor acts being escalated into serious criminal accusations. Like the days of yesterday, it was a way to ensure availability of a cheap labor force and gain wealth after slavery was abolished at the end of the American Civil War. The unresolved situation of freed slaves goes on; not really being free. 19

Similarly, in some schools with Black and some Latinx children, the least infraction or acting out can lead to suspensions, more suspension and eventual expulsion. If there is no one at home to supervise or direct the suspended or expelled youth, some suspended and expelled students may look for a place to be welcomed and feel a sense of belonging. 20 Eventually many of the African American and Latinx children end up in prison.

Often the youth has experienced growing up with violence in the home or community without direction and lose their sense of self-worth and self-esteem. Some school teachers and administrators make rules whereby intolerance of any type of acting out behavior, results in a Black child to falling into the disposable category. They do not want to take time to understand what or where the youth is coming from or know when they are asking for help. Therefore, a youth can easily want to be identified with a family who takes an interest in them as this is a basic human need. The gang family is where this can end up after feeling trapped and not seeing a way out of loneliness and poverty. Thus, some school systems serve as the pipeline to prisons. 21

To guarantee a way to fill up jails and prisons, poor Black and Latinx communities experience what I call "the hunt." In other words, those on probation are watched or "spot-lighted" like a hawk by local police, waiting to jump in for the "kill." No pun intended except it can often end up in a killing. Spotlighting brings in bounty by the dozens.

When a trial or hearing comes around, with lack of resources to have a lawyer, lack of understanding for the legal procedure and no ally or community support that can speak up on behalf of the arrestee, prosecutors can do what they want to an accused violator, juvenile or an adult. This incarceration does not resolve the problem and it's the poor Black or Latinx, often the targets of the spotlights, who end up in cages. It does however fill the bank accounts of the private prison industries. Profit is made from a myriad of things to include supplying accounts as a liaison with family of incarcerated persons to provide the incarcerated relative or relations-greeting cards for holidays and birthdays, sandwich coupons and other services such as phone calls. Ankle bracelets are provided when released and the list goes on. 22

According to PEW, Blacks and Hispanics make up larger shares of prisoners than of other U.S. populations. Supporting these facts, World Atlas reveals,

"The US prison system is the main source of punishment and rehabilitation for criminal offenses. The system incarcerates more people than any other country in the world." The United States has well over 2 million prisoners and China comes in second with 1.5 million, but China's incarceration rate is only 118 per 100,0000 people.

World Atlas also reports that the US, El Salvador, and Turkmenistan have the three highest rates of incarceration in the world, for varying reasons.

El Salvador has many people in prison due to its ongoing gang crisis, and Turkmenistan puts people in prison for objecting to the government and following the wrong religion.

Research has shown that putting a high percentage of a nation's population in prison has a negative impact on society as a whole, perpetuating a cycle of crime and violence that is generational. 23

According to Stanford Center on Poverty and Inequity, "racial inequity exists in the US Criminal Justice System. Black persons are imprisoned at more than five times the rate of whites; one in ten black children has a parent behind bars, compared with about 1 in 60 white kids." 24 Is this an accepted norm since this people-disposal system crisis has continued so long?

Transitioning to another aspect of our disposable world story, let's move into what most people consider comfortable talking about and sharing with others. That is "Let's save the planet and reduce greenhouse gases in the atmosphere." Clubs, alliances, businesses and organizations have formed for this cause. It is important that these groups include the people who are most affected by climate change. Kudos to those who are finding equitable solutions for all people involved.

Starting with the trash conversation, what about this big one called plastic waste? This is what affects environmental quality and ecosystem health. The amount of plastic waste generated in the United States' estimated to enter the coastal environment in 2016 was up to five times larger than that estimated in 2010, rendering the United States contribution among the highest in the world.

Plastic can be found everywhere. This includes bags, sacks, wraps, packaging, bottles, jars and containers. Plastic is found in durable goods such as appliances, furniture, casings of lead acid batteries and other products.

Plastic can be considered as one of the most important materials in human history since it first came into commercial use more than 50 years ago. It is estimated that the United States is the world's largest generator of plastic waste with the average American generating roughly 130 kg of plastic per year. About 12% of the 292 million tons is attributed to the United States. Many items are improperly disposed of and are contaminated. And the fact is that not all recycling facilities can treat every type of plastic product collected. Therefore, a large amount of plastic waste that could be recycled, ends up in the landfill or incinerated for energy recovery. 25

Now, I know you're not surprised when you realize that the United States export huge volumes of scrap plastic to developing countries every year. It's supposedly a convenient and cheap way of dealing with the United States waste. However, many countries where this is exported to have inadequate waste management infrastructure.

Once upon a time, the United States would ship the majority of plastic waste to China. That was stopped due to restrictions on waste imports. Shipments of United States waste goes all around the world; and lately, Canada has become the top market for United States scrap plastic. The US exports millions of pounds of plastic scrap to developing countries such as Malaysia, Vietnam and Indonesia. An issue with that exporting structure is that a majority of plastics are not biodegradable. Thus, plastic can stay in the environment for hundreds of years. And we know what happens with our oceans as they become polluted with plastic. 26

When we dispose of our waste materials by burying it and covering it over with soil, we create gases. Some say that methane is over 25 times more detrimental to the atmosphere than carbon dioxide. So, CO_2 can't take all the blame. Landfills are the third largest source of methane in the US. Thus, landfill gases have an influence on climate change, containing both CO_2 and CH_4. 27

Methane is a potent greenhouse gas 28 to 36 times more effective than CO_2 at trapping heat in the atmosphere. 28 The United

Nations Environment Programme (UNEP)'s Global Methane Assessment reports that "Cutting methane emissions by 40 to 45% by 2030 would almost immediately slow the rate of global warming, and shave 0.3 degrees off of the increase in global temperature by 2040." 29

Plastic Waste

Can we also cut down on our use of plastic? We buy items in plastic and throw them in the garbage and think nothing more of it. When you take your garbage to the landfill what happens? The food waste and other organic matters are broken down by bacteria that do not need oxygen. These food particles and garbage return into the air in the form of methane which is an extremely potent greenhouse gas. In some countries garbage and waste is burned which release carbon dioxide and even more gases.

Incineration is on the rise in parts of Europe. In England, nearly 11 million tons of waste were burned at waste-to-energy plants, up 665,000 tons from a previous fiscal year. The facilities are designed to contain emissions, and the practice has sparked strong reactions both for and against, among environmentalists and scientists. However, a recent study by the nonprofit Zero Waste Europe found that even state-of-the-art incinerators can emit dioxins *[a highly toxic compound produced as a byproduct in some manufacturing processes, notably herbicide production and paper bleaching.]* This is a serious and persistent environmental and harmful pollutant. 30

ACTION

THE NEW STORY OF HEALING & SOLUTIONS

Every solution and resource births a new story for change. If we want to see changes, we must take action.

In this part of the chapter are possible solutions and resources to help inspire actions so we can have opportunities to live new or revived stories to:

A) Awaken to the reality that when we speak about a disposable world, we are speaking about the disposal of people and trash/garbage.

B) Find ways of helping disposed people feel and be whole, accepted and inclusive in society.

C) Find an action item as a pathway towards a climate career, social justice career, research or volunteer project that is just waiting for you to choose from. That is, unless you have already been triggered into action.

1. **HOMELESS PERSONS**: possible items to pursue
For those who want to pursue a boots-on-the-ground approach- If you don't know what to do when you approach a homeless elder or other homeless person and you want to do something:

- First of all, be sure that you are safe. You can't help anyone if you put yourself at risk of being harmed. Also, it is advisable to wear a mask and gloves during pandemic times.

- Treat all homeless persons with respect and speak to them. Have a conversation, if allowed. Give a smile and let them know you do not consider them as garbage.

There are lots of causes and a because as to why they are homeless.

- If a conversation is held ask them their name and tell them yours or how you rather be called, as an exchange rather than an outright inquisition of questions. The key thing to know is if they have places to go every day or certain days for food and if they have shelter and warmth at night. If one is non-compliant, move on do not agitate them.

- Know that 211 is a # they can dial provided they have a cell phone *(most do, if coherent, as they either use their disability check or other ways to obtain them- I do not know how phones are paid for or provided by).*

- When it is chilly and cold, and you know if someone sleeps in their vehicle or in a tent, you can offer them disposable "hand-warmers." These can be purchased at sporting goods stores. These keep their feet and hands warm. It gets cold on the Monterey Peninsula near where I live, but it is not like places that are much colder with lots of ice and snow and the problems associated with homelessness are graver.

- Become an advocate to someone you have connected with who needs to have help in making phone calls, keeping schedules, appointments, etc. and seeking the help they deserve in order to function and thrive. Use the Homeless Services Resource Guide mentioned above or suggestions from action-oriented individuals.

ADDITIONAL SOLUTIONS FOR THE HOMELESS:

- Connect with various non-profits in your area to be fiscal sponsors or sponsors to assist in getting funding to have built or purchase **small houses** with supervised staff on a property. Propose open ground that a city or rural area might approve of. In Monterey County suggestions could be the old Fort Ord. The "not in my backyard" (NIMBY) problem is that home owners and residents in the affluent County of Monterey might resist. However, in Monterey County with a coalition of folks as allies and non-profit funding, there is a possibility.

- Check out what is being done in your County or municipality. There might be a Coalition of Homeless Service Directory that is available for housing solutions.

- In Monterey County, search: The Coalition of Homeless Service Providers and be guided according to what you need. 31

Please note that there are volunteer opportunities. One example of how to volunteer is to search Community Human Services with an online volunteer application available here or

Contact:
(831) 658-3811 and receive one by email
Also search for similar titles as the ones listed here for Homeless & Affordable Housing Resources in (Monterey County in California):

Housing First Approach - Housing First is a homeless assistance approach that prioritizes providing permanent housing to people experiencing homelessness, thus ending their homelessness and serving as a platform from which they can pursue personal goals and improve their quality of life.

Project Home key - Project Home key is an opportunity for state, regional, and local public entities to develop a broad range of housing types, including but not limited to hotels, motels, hostels, single family homes and multifamily apartments, adult residential facilities, and manufactured housing, and to convert commercial properties and other existing buildings to Permanent or Interim Housing for people experiencing homelessness.

Coalition of Homeless Service Providers - The Coalition of Homeless Services Providers is a 501(c)3 nonprofit organization that serves as Monterey and San Benito County's Continuum of Care (CoC) lead. In the most basic terms, CHSP coordinates services among partner agencies, ensure our community is following federal guidelines, reports homeless statistics to federal and state governments and seeks funding for the homeless services community.

Community Human Services - Community Human Services is a nonprofit agency dedicated to providing high quality mental health, substance abuse and homeless services to Monterey County residents to help them reach their full potential.

Types of Low-Income Housing Programs - The Coalition of Homeless Service Providers has an extensive list of housing options at this link. Please note, due to a shortage of funding as well as limited shelter beds available, these programs may not always have capacity to assist someone in need of housing.

I-HELP Monterey Bay - Interfaith Homeless Emergency Lodging Program is operated by Outreach Unlimited, a nonprofit 501(c)(3) organization in the Monterey Bay offering short-term lodging and evening meals on a nightly basis.

Front Porch - Front Porch is a nonprofit that specializes in senior living communities and programs. Front Porch helps individuals fortify their own health and purpose by building strong and engaging communities, connecting people with the services and relationships they need to thrive, and providing compassionate support. On April 1, 2021, Covia and Front Porch merged to become one organization under the Front Porch name.

City of Monterey Affordable Rentals - Located within the City of Monterey are a number of rental apartment units that are required by affordable housing agreements to be rented at prices that are affordable for low and moderate-income households.

ECHO Housing - ECHO›s ultimate goal is to promote equal access to housing and provide support services which would aid in the prevention of homelessness and promote permanent housing conditions.

2. **RESTORATIVE JUSTICE** for juveniles, prisoners or those at-risk

A source for helping those who have committed a crime and those who might potentially be at-risk of doing so is Restorative Justice. To restore means to bring back and repair. The term Restorative Justice refers to a system of justice that focuses on the rehabilitation through a meeting with the victim and the perpetrator, sometimes having community supporters and definitely a mediator or facilitator.

Since the criminal justice system can sometimes be complicated to those who have been harmed and are experiencing trauma, many victims want the rehabilitative solution rather than the punitive approach by most criminal justice systems. Restorative justice lends itself to facilitate the healing of all impacted parties. It gives the victim an opportunity to communicate with the perpetrator and express how they've been affected by the crime. It also gives the victim a chance to have any un-answered questions resolved. Of course, both parties have to agree to take part with any party stopping the process at any time if they are uncomfortable with it or

if they just need a break. This meeting is completely option-
al and is not about passing judgment. It is about repairing
harm. Everyone has the ability to change. Restorative Justice
helps people to move on with their lives.

In studying various forms of restorative Justice, they all
conclude that there is a high rate of success in reducing repeat offenses. Perpetrators or offenders have the chance to
make things right. This is a learning process that very often
leads to a crime-free lifestyle. As harm has been repaired,
recidivism rates reduce. 32

There are different types of Restorative Justice formats
or circles where emotional harm needs to be healed, where
"spirit is gone" as some Native Americans might put it and
repair is needed. They work in many settings- schools, communities, juvenile detention centers and other criminal set-
tings as well as prisons. To mention other types:

- **The Community Building Circle-** Organizations and
  communities get on the same page and move ahead with
  healing and working together.

- **Harm/Conflict Circles-** In some schools where there is
  extreme violence in hallways where students are actually
  fighting these conflict circles are needed. An example to
  alleviate continuous suspensions leading to loitering and
  gang behavior, when one is allowed back at school for a
  certain number of days. With expulsions, in some cases,
  there is hardly any chance of connecting with productive
  members of society.

- **Victim-Offender Mediation** –Victim and Perpetrator
  meet face to face with a mediator and possibly someone
  from each community, if desired. This gives support to
  either party.

- **Family Group Conferencing-** where victim, offender,
  family, friends, supporters and community members

come together and discuss what happened. There may be an agreement whereby the offender pays for damages, works, re-build, or whatever else satisfies the victim. 33

Even in what one might consider a primitive culture, this same type of restoration can be found in varying forms. In certain regions in South Africa, Ubuntu is used when someone does something wrong. This person is taken to the center of the village and surrounded by his tribe for two days while they speak of all the good in this person, sometimes realizing that mistakes are made, which is really a cry for help. They unite in this ritual to encourage the person to reconnect with their true nature. The belief is that unity and affirmation have more power to change behavior than shame and punishment. This is known as Ubuntu-humanity towards others.
Politicians such as Nelson Mandala and the general public use this catch-all term "Ubuntu" referring to the country's spirit of togetherness, ability to work together towards a common goal as examples of collective humanity. 34

*Contact:
  A) Search by Your State's Restorative Justice programs.
     Example: California Restorative Justice in the Schools or Restorative Justice for Juveniles in California.

  B) Search: Restorative Justice or
     https://restorativejustice.org/where/ 35

3. More Action Items for Youth and Communities

  A) Racial Healing Tool Kit Activities for Youth Involvement 36

  B) Explanations & Suggestions to Dismantle School-to-Prison Pipeline - YouTube Search: What is the School-to-Prison Pipeline?

Another Solution of transformation for those considered disposable people, is found in Chapter 7, Part II, Disposable People-The Enneagram Solution Story, Unlocking Your Inner Prison

LOS GATOS, CA 95031
1 408 600 0074
INFO@ENNEAGRAMPRISONPROJECT.ORG

4. DISPOSABLE TRASH & GARBAGE possibilities

According to PEW Research, plastic waste is entering the ocean at a rate of about <u>11 million metric tons</u> a year, where it is harming marine life and damaging habitats. «Breaking the Plastic Wave,» a global analysis using first-of-its kind modeling, shows that we can cut annual flows of plastic into the ocean by about 80% in the next 20 years by applying existing solutions and technologies.

1) REDUCE growth in plastic consumption to avoid nearly one-third of projected plastic waste generation by 2040

2) SUBSTITUTE plastic with paper and compostable materials, switching one-sixth of projected plastic waste generation by 2040

3) DESIGN products and packaging for recycling to expand the share of economically recyclable plastic from an estimated 22% today to 54% by 2040

4) SCALE UP COLLECTION rates in middle-/low-income countries to at least 90% in urban areas and 50% in rural areas by 2040

5) DOUBLE MECHANICAL RECYCLING capacity globally to 86 million metric tons per year by 2040

6) DEVELOP PLASTIC-TO-PLASTIC CONVERSION potentially to a global capacity of up to 13 million metric tons per year*

7) DISPOSE securely the 23% of plastic that still cannot be economically recycled

8) REDUCE WASTE EXPORTS into countries with low collection and high leakage rates by 90% by 2040

TO LEARN MORE ABOUT THIS TO DETERMINE HOW YOU CAN HELP & OBTAIN COPIES OF THE PEW RESEARCH MATERIALS.

A) Search: Breaking the Plastic Wave
   Search Breaking the Plastic Wave 38

B) The PEW Charitable Trusts: PreventingOceanPlastics@ pewtrusts.org or Search Preventing Ocean Plastics 39

C) SYSTEMIQ: OceanPlastics@systemiq.earth or search Ocean Plastics 40

ADDITIONAL DISPOSABLE TRASH & GARBAGE Info

What we recycle is important. How we sort is even more important. In some states there are specific guidelines for even recycling food scraps? It is encouraged that climate action groups can gather and ask legislatures to pass laws to help reduce greenhouse gases in landfills. A California Senate Bill is just an example of this called SB1383 that came into effect January 1, 2022. This State law requires residents and businesses to keep compostable and recyclable materials out of California's landfills.
Search: SB 1383 Education and Outreach Resources 41

The California SB1383 Climate bill is intended to reduce methane emissions that are generated from landfill and to divert organic material from landfill to be composted. Requirements and timing of the law are:

2022: • Provide organics collection service to all residents and businesses

• Establish an edible food recovery program

• Conduct education and outreach

• Procurement (of resulting compost)

• Capacity Planning

2024: Jan 1, jurisdictions must take action against non-compliant entities

Requirements are all programmatic in nature. Processes and materials need development into programs.

However, it won't be until 2024 when jurisdictions must take action on non-compliant entities (those not collecting food waste for compost or recovering edible food)

The County climate action group that the author is a member of called CSMC (Communities for a Sustainable Monterey County) is planning how to help implement SB1383 with the help of all its local area affiliates.

FOR MORE INFORMATION & SUGGESTIONS AS TO HOW YOU MIGHT GET INVOLVED IN YOUR AREA.

Search: California Department of Resources Recycling and Recovery (CalRecycle) 42
Additional Resources: The author is continuing her research on this topic and can provide you with additional solutions beyond

the publication date of this book. Just email info@lavernemcleod.com to request this.

Let's not forget that any of these actions will also help:
- the winged ones
- the crawlers
- the swimmers
- the hoofed and claw-toed ones
- the microscopic ones

Chapter 7

Part II
Disposable World

The Enneagram Solution Story
Unlocking Your Inner Prison

This chapter can be thought of an addendum to Chapter six's Disposable World or an addendum to living. There is such a vast amount of research materials on climate justice as people want to "save the planet." However, fewer stories emphasize saving the people on the planet that are vulnerable, constantly discriminated against and judged to be worthless and therefore disposable. This chapter focuses on imprisoned human beings and finding a pathway to freedom.

Does your mind ever go to the place of wandering what might be happening to people behind bars, in cages of captivity, called prisoners? The flip side of prisons are those who think of themselves as free and yet live in psychological bubbles of confinement as well. Does one know their personality, defined by their character, temperament or disposition, well enough to distinguish whether they are confined or free?

The unique story to being free from certain confinements derives from the Enneagram. According to Riso and Hudson, its origin takes us from its ancient roots dating back some 2,500 years ago to the modern enneagram of personality types and does not come from any single source. As one studies the Enneagram history, they will discover the symbol with the nine personality types that comprise this powerful learning entity. 1

Constructed from this knowledge, Susan Olesek created and founded a program called EPP or Enneagram Prison Project. It is a non-profit dedicated to serving the incarcerated and their potential future. The Enneagram is quoted as "a psychological system brought to Stanford University in the 1990s, is an incisive self-awareness tool which identifies the unconscious cognitive, emotional, and behavioral strategies which underlie virtually everything we do. Taught worldwide to individuals, business leaders, spiritual directors and the incarcerated, EPP brings the Enneagram to those whom they feel may need it the most. By reflecting the light seen in people back to themselves, this illuminates a path one can reliably follow. People are empowered to find their way back to their own innate intelligence and wisdom, to re-member their best selves."[2]

Subsequently, this intersection of reflecting light on prisoners is where a lot of us so-called "free" and "never been incarcerated persons" make judgements about those subjected to prison cells. I admit that I was one of those "free" persons who had conditions by which I made judgements. I told myself that if it was just only minor drugs or racial targeting that got them in there, I could be ok with them. My viewpoint changed rather quickly after hearing from former incarcerated persons about major crimes that got them locked up and the power of their speaking about their Enneagram types in depth. This helped me understand my own self-made imprisonment. This is how my personal story about the Enneagram began:

I was writing this particular book at the time reflecting on climate and social justice. I had been amazed thus far as to how the Universe provided me with material for it, almost always unexpectantly. The particular chapter called *Disposable World* was unfinished. Little did I know that the word "enneagram" with its programs would be just what I wanted as an excellent solution for people considered disposable by society.

I first heard of Enneagram, pronounced "ANY-a-gram" when I moved to Esalen Institute in Big Sur, CA in 1979. I said that one day I would dive into its significance. That day happened 42 years later, late October 2021 when I first actually listened to someone speak about it. I used to avoid what I thought was labeling me. After all, being an African American with many labels and projections, no one was going to assign me a number like cattle or chattel slavery. I was on Esalen property that day for a documentary and happened to be looking for a place to eat my lunch where I felt psychically comfortable and was outdoors. I chose to sit with three friendly souls who welcomed me to sit with them. For privacy reasons, I am calling them-Allen, Celia and Carol. As the friendly conversations began that "Enneagram" word popped up and I learned that they were attendees in an Enneagram workshop that week referred to as the Enneagram Prison Project (EPP). Curious about the project, I inquired more. I was so fascinated and said, 'This is done in prisons?' I had to pause at this and eat more lunch.

Serendipitously, the phrase Human Potential was mentioned by one of these friendly souls and I was even more stunned because the person who had just interviewed me for comments in an Esalen documentary asked me about the Esalen Human Potential Movement. I kept listening to the lunch table talk and especially how EPP helps a person reach their full human potential.

What went through my head was what the Esalen interviewer spoke about and asked me earlier that same day:

Sometimes when we talk about Human Potential, and the history of Human Potential at Esalen, it feels like there has been a significant blind spot, in that, in a sense, it feels like the privilege to search for one's human potential was bestowed only to a certain demographic-white, middle class and so on. What are your thoughts, having lived through this era?

After responding to the interviewer's question, it made me think deeper about what is called the Human Potential Movement. I felt so thankful to hear about the inclusivity of EPP, that someone actually sees prisoners as human beings and not human waste to be disposed of and exited from society.

From this approach to Human Potential, I was not concerned about something as trivial as labeling me with a number. This Human Potential work was and is like the hands of angels helping some prisoners to lift themselves up from a world where they were not expected to succeed, disposed of like food waste, to rot in cages without feeling any self-worth and die without dignity.

Having lived at Esalen for three years and having experienced the land where often magical transformations happen, I felt more attunement to the word, "Enneagram" this time around. Being excited and thankful that this information resonated on a heart drum that I could beat, I wanted to learn more.

I didn't think twice about asking permission to place this information about EPP in my book to reach others. Carol went to another table where other EPP seminarians were seated to ask who specifically, should I contact for this. I was told that Susan Olesek would be the one to ask. I was given her contact email. This ended in a zoom call where I was granted permission and also was given information about a program called 9Prisons ONE Key (9P1K), open to anyone. That way I could learn more about how the Enneagram works and realize its personal value. I was sent information and enrolled in 9P1K, an 8-week intensive program. These are some of the things that I absorbed from the course.

First off, this course is nothing to be playing around with. The commitment and seriousness from point one is evident with the intensity of the questions that one must answer prior to being accepted in the course. That is, after acknowledging your interest to take it.

Secondly, 9P1K requires checking into yourself, not just at the outset of every weekly session. It can happen when you are alone or with others, gauging yourself. This comes from a deep plunge into the homework or weekly focus outside of class time that often takes 2 ½ hours to complete. This gave me a lot to think about as a personality type 8. Being impressed with the various personality types exemplified by Guides, Ambassadors and participants, I particularly gained a lot when it was time for all 8's to respond to the questions of the Guides. I truly felt the peace of mind expressing myself fully with a lot more confidence as I did not have to tone myself down. That felt good!

Furthermore, 9P1K awakened me to pause when my inner alarm bell warns me before, and shortly after in some cases, that I am thrashing out gut-evoked words that I might regret later. These are words that could wither one's spirit and crush the goodness of what could have evolved, if done in another way. It's like me observing me saying, 'Hello LaVerne, this is who you are, get comfortable and become a better you.'

At mid-point through 9P1K, I absorbed that it took a lot of wisdom, fearlessness and love to create such powerful courses from the Enneagram. I consider Susan Olesek a "visionary angel." I am saying this because she saw and continually sees the good that each human possesses which may be locked inside themselves. Because of her compassionate, unbending kind-heartedness, she is a true Human Potentialist. She and her trained staff bring light to human beings. They are the catalyst to help humans dig deeper and evolve at their willingness to shine.

Additionally, I took in or assimilated an enriched experience that is shared in this course coming from formerly incarcerated persons called "Ambassadors." I say the Ambassadors were saved, redeemed, rehabilitated and healed because they chose to put years of participating and engaging in becoming transformed human beings by working in-person with the Enneagram Prison Project (EPP). They are the ground-busters and root of 9P1K's virtual course. There is so much I learned and am still learning from the Human Potential models (Ambassadors). Their stories have given me a sense of understanding of just how effective EPP really is.

Beyond my expectations, one such model Ambassador struck my compassion chord with his depth and credible truth-telling that seemingly came from his core. I instantly wanted to know more about his quality of experience that unlocked for him a path to emotional and physical freedom. I felt that he could be my 'brother.' For this means, a kindred spirit-brother of the 'don't let the chains hold you down' member, embedded in the progressive African-American psyche or soul. This Ambassador, whom for privacy reasons, I will call Dwayne, was interviewed by me. First, I sought permission for this conversation with Dwayne from Susan Olesek, EPP and 9P1K founder, and of course from Dwayne. The following is what came from that conversational inquiry.

Interview with Dwayne

With a 25 to Life sentence hanging over Dwayne's head at age 39, he had to choose between losing or gaining. Dwayne stated, what got him this term. "I was a drug addict for one. I became addicted to drugs in the early 80's and I had 3 strikes of 25 years and I took a deal of 25 years to life. I said I can do 25 years but I can't do 150. I took responsibility but I didn't know how to make changes in my life."

Since Dwayne is now a free man, off parole after serving 18 years at San Quentin State Prison in California, I asked:

What was the Driving Force behind you not giving up while incarcerated that led you forward?

"My first main driving force was seeing that this is not where I wanted to see my life to end. I had people that were praying for me and in my corner and that helped. My grandmother's love for me, her support for me and her words stuck to me. She said, 'boy ain't that 25 to life hit you yet? Isn't that enough to bring you home, where you know is good?' She has been really my rock through all of this. I think that was my driving force for my change."

Other forces were working in Dwayne's life to see the visual reality of freedom while incarcerated. He expressed, "I still had doubt. I didn't think I was smart enough, that I could do certain things. My first cell mate at San Quentin was a lifer also. And after a couple of weeks of us being in the cell together, he seeing that I was not much into gangs and not into non-sense. He came in from work one night and asked me, 'what do you want to do? What do you want to do with your life? Is this where you want to be?' I said NO and he sought me out on the yard the next day. I don't know if any of you have ever been to San Quentin. There's a lower yard at San Quinton and there's an upper yard. You can look out across the top of the fence and you can see out into freedom. You can see Marin County. You can see the Bay. You can see cars, everything. And I looked out into the lower yard and that was another significant moment in my life where I said that I am through. I am done.

This is it! I went back in and I told him this and I cried. He sat there with me and allowed me to have that space and said, 'Well, I got you. There is something more than this and you can do it. I am going to help you find a way, for you to get yourself together so that you can get on the path to freedom.'

"I took responsibility the second time I went to court. I told them that I did the crime and was not trying to do 150 years either. I asked the judge to ask the district attorney to offer me a deal that I could take, going to court, having kids at the age of 15 and 16. Having them witnessing me going to court and being in the newspaper was too much for me."

"It took time to mature because in my early times, I didn't get it. It wasn't clicking. My drug addiction, coming in and out of prison on violations on different cases just wasn't clicking." Thus, Dwayne had to move forward from what he said some inmates were doing. "a lot of African Americans in prison blame the system. They blame this and they blame that and they don't look at themselves as part of the problem and I did."

"I started doing different groups that included Alcoholics Anonymous and Narcotics Anonymous and one of my biggest changes was when I started going to college. I hadn't been in school since I was 17 years old and got my GED. I started going back to school in 2008. I want to say at age 42 or 43 I started going back to school and I was scared. I didn't think I could do it. I had to start at the bottom. I had to start with fractions and decimals and algebra was not easy. I did not know how to write a thesis statement. I didn't know what a topic sentence was. I didn't know how to organize anything like that and I was taught that. I got my college degree."

And with all of this to prompt forward momentum, "I continued taking classes in groups and EPP courses. It took me 2 years to complete the course. The most difficult part about doing EPP was being willing to go back in the past and look at my hurt. Looking at what and who hurt me emotionally. I had to break down a wall that I built around myself for my safety or so I thought.

In the process, I was destroying myself. That was the difficult part to face. Just looking at the background I had with growing up,

I understand my choice. Once I was able to start taking classes I really understand that this was my responsibility. This was nobody else's responsibility. Things happened in my life that I held onto or I don't know the exact word but I had some stuff that went on in life that I used to hold onto to that I didn't know that I was holding on to when I first came to EPP groups. It took these classes for me to be open to childhood psychology or early development of childhood abuse. And the easy part was the people coming in and explaining and giving their heart freely and making me feel like a human being. Showing love in a dark place. EPP is what it took for me to get and understand about myself that I was able to say, I got it."

I asked Dwayne *"What did you get or gain from EPP?*

He further shared,

"My childhood does not have to hold onto me today but it did curb me into the choices I made but I don't have to hold onto that anymore." He reiterated that "I lived in a box. I had built a prison for myself inside my head that I wasn't going to allow anybody to hurt me. I wouldn't allow anybody in to help me either."

Within the time-span of Dwayne's EPP classes, he progressed to "mature and having self-will and self-acceptance that helped a lot.

He further stated, "San Quentin was a Blessing because of the area and everything and everybody that comes in. He says about EPP instructors called Guides that came into prisons: "You guys bring heart into a place that is heartless." Listening to Dwayne, it seemed that a true reality slogan that could be EPP's brand.

Dwayne seemed to radiate light as he was eagerly expressing himself about EPP. "There were many people along the way that helped me change my thinking, open my mind and EPP was the last key that I needed. I was holding onto a grudge with a stepfather that I felt tried to kill my Mother. I was holding onto a grudge that a father who had passed away but tried to kill my Mother and me while I was still in her. So, I was carrying all of this and nobody ever explained to me that this was not my fault. That this was never about me. My biggest issue that I was holding onto was that people were covering up for somebody that had hurt my Mother.

And I would not accept it and I was just letting that eat me up. And once I realized that and also realizing what people do and don't do because of how they are taught or the education they might have. If you don't know better, you can't do better and once I accepted that in myself, I had to accept that in everybody else. I had to look at everybody, even in my past, I had to look at it and they didn't know better. In the era that they grew up in that was what they were taught and that's how life was. I don't hold that grudge anymore. Yeah, I have some things that I still work with but I don't hold grudges anymore. I try to find a way to process and be better." And with these types of openings that unlocked Dwayne from his emotional prison, I asked him another question.

What's most important to you right now in your life?

"My FREEDOM. I am finally free mentally and physically. I never want to subject myself to drugs and alcohol because the prison was just a by-product. I used to try to escape. I was escaping through drugs and alcohol and I really didn't like alcohol because both my dad and my stepfather were alcoholics. It was drugs and peer pressure that got me going. I am mentally free and I am physically free and I have the paperwork to prove that I am physically free."

What's it like being free? Going where you want to go and doing what you want to do. How does that feel?

"It feels good but my life is still in the process like I said, I made a choice to come home to be closer to my family and siblings. My brother hasn't left me even though we are not biologically connected. We grew up together and he has not left my side but we are working on growing. I am trying to save money so I can get in a better position. I really haven't gone anywhere yet but I can go at any drop of a hat if I want to. That's a great feeling. I can get in my vehicle and go as far as it takes me and that's the Blessing right there."

And I see that it is a Blessing that you can share that with other people and that you are a part of the EPP process and can learn from your experiences. This leads to one last question I want to ask you:

Where do you see yourself five years from now or do you even go that far?

"I hope to own my own property within that time. I want to own my home. My credit line right now is better than I ever could imagine. It's just about building capital, finding a way to make enough money to keep away or put away so I can buy my own piece of land. My brother, an x-felon who has been out for like 20 years, is teaching me this re-modeling and building stuff and its great. But my goal 5 years from now is to own my own home. He builds homes, he has homes, he has apartments and it's about building longevity for his kids. Once I can have my own home and I can have a place for my grandkids to come and say, 'I am at Grandpa's.' That's all that matters."

This interview concluded with gratefulness and accolades for Dwayne's sharing. Both the EPP Community Weaver and myself could see the vibrancy and radiation Dwayne was emitting. We shared that with him and encouraged him to keep that light shining that projected peace and accomplishment.

I recommend to anyone who truly wants to know themselves to do so to register in the 9P1K course. Each person will have a different experience and different take-aways. Prior to knowing if I was actually accepted to participate in the course, I indicated that I am open to see what could come forth for me. I had no idea that I had made a prison for myself. I had blocked out acknowledging all of me that was in my face all the time. I now feel more confident and express myself with greater ease.

It is this type of program that everyone from adolescents to adult can benefit from to truly know themselves. As quoted from the ancient Greek poet, Pindar, "Learn what you are and be such."

Do so by unlocking yourself from your inner prison. Here is the KEY:

Register for a 9P1K course. https://enneagramprisonproject. org/9prisonsonekey-training/

Or Search: 9P1K or 9PrisonsONEKey

Or email: info@enneagramprisonproject.org

Or Call (408) 600-0074

Become a fundraiser to keep the organization going to spread the human potential wings using the contact info above. Or

Write:
Enneagram Prison Project
P.O. Box #804
Los Gatos, CA 95031

How do you feel after reading this chapter?

Chapter 8

Deforestation

This chapter starts with an ode which is also a song written by the author.

Ode to Deforestation
A Song of Sadness

Deforestation, like defamation to nature
To the spirit of life

Cutting down trees
Upsetting the bees,
The bugs, the growls, tweet, sizzles, ruffling wings and squeaks

Invasion, exclusion
Is death the conclusion
For mankind's greed
to cut down beautiful and ancient trees?

With a mindset for lack of compassion
Brings forth a lot of reaction
To take homes and food from the vulnerable people
The animals and other creatures

When will the pillage and raids ever stop?
Have we turned back histories clock?
Where the harmed
Have to run
Die

Survive
Or cry
Does this make mankind so great?
To commit extinction rape?

©LaVerne McLeod 5/13/2022

Forest Tree Trunks/Deforestation

Of course, we all know that deforestation is defined as cutting down and clearing land masses, trees, and essential habitats on a massive scale. And we know that the "I want and I take it no matter who or what it hurts" attitude prevails. Historically, mankind has taken what it wants for profit. Will this mindset ever change?

In this chapter, issues of cutting down life forces that help people thrive and revive on Earth are dealt with in two ways. What happens to the climate when we destroy our forests? What happens to people affected by this damage, physically, emotionally, economically and metaphorically?

Let's consider a few facts first. According to National Geographic, about 17 percent of the Amazonian rainforest has been destroyed over the last 50 years. Although vanishing quickly, these forests cover about 30 percent of the world's land area. 1

Since 1990, the world has lost 420 million hectares or about a billion acres of forest, according to the Food and Agriculture Organization of the United Nations-mainly in Africa and South America. It also reports that many different species of plants and animals reside in the soil, understory and canopy and are lost. Estimates of the total number of species on Earth range from 3 million to 100 million. 2

The destruction of the rainforests is the destruction of one of our richest ecosystems. Trees absorb the carbon dioxide that humans exhale as well as absorb the heat-trapping greenhouse gases released by human activities. To get an idea of this destruction area, according to Yale Environmental report, the Brazilian Amazon is the size of Western Europe, an area larger than France has been cleared. 3

Mankind's greed for the forest for profit creates a lot of damage for the climate as well for the people relying on the forests for sustenance. What mankind is doing is taking land for agriculture, cattle raising, logging and building homes. Agricultural crops such as soybeans, palm trees and cocoa are the by-products of deforestation. In Malaysia and Indonesia, forests are cut down to produce palm oil, used in shampoos and in foods.

Cutting down or burning forest accounts for 80% of deforestation. With 200 million head of cattle in Brazil's Amazon, it has become the largest exporter in the world, supplying about one quarter of the world market. Providing the world with wood and paper products accounts for the felling of numerous trees every year. Oftentimes, loggers build illegal roads to acquire access, thus cutting more trees. The wood is also used as home building for what is called the "urban sprawl" or the spreading out or development of suburban homes near cities.

Therein lies the problem. 4

The consequences are so grave because with deforestation, the trees are no longer there to absorb carbon dioxide which fuels earth's warming trends. Thus, the speed in which these gases enter the atmosphere is increased. Bare ground that is left exposed after cutting down forest often reflect more sunlight than the downed trees which created shade and cooling. Carbon dioxide, nitrous oxide and methane are released with exposed stripped soil. In Indonesia's swampy lowland area where forests are cut down, water-logged peat dries out and decays, carbon dioxide is released when the peat is burned by man.

The social justice side of deforestation crushes villages of people. Let's take Indonesia for example. Other than cutting down trees from the swampy lowlands, the swamps are drained and the remaining brush is burned to totally clear the land. These swamplands or peatlands are made up of nutrient-rich soil from thousands of years of decaying plant matter. So, when the peatlands are burned, they release a thick, noxious haze that spreads throughout the surrounding areas. This haze travels to neighboring islands and countries on air currents. The fumes poison wildlife and rural villages. It is reported by the Borgen Project that addresses deforestation in Indonesia, that more than 100,000 annual deaths can be attributed to the inhalation of particulate matter from the burning peatland fires. 5

In the South American rainforest regions, water cycles can often get dirty from soybean farmers and beef ranchers. This can have devastating effects on coffee crops by jeopardizing the viability of different varieties of coffee. 6

ACTION

THE NEW STORY OF HEALING & SOLUTIONS

Every solution and resource births a new story for change. If we want to see changes, we must take action.

In this part of the chapter are possible solutions and resources to help inspire actions so we can have opportunities to live new or revived stories to:

A) Waken to the consequences of deforestation and find

ways to solve the problems it present

B) Find ways to help people affected by deforestation

C) Find any action item as a pathway towards a climate
   career, social justice career, research or volunteer
   project that is just waiting for you to choose from. That
   is, unless you have already been triggered into action.

Action Items to Pursue:

1 **Manage Forest Resources-** Get involved with Forest
   Service International Programs that work with forests in
   Indonesia as partnerships: International Programs | Main
   phone number +1-202-644-4600 | Fax number +1-202-
   644-4603 | 1 Thomas Circle NW, Suite 400 | Washing-
   ton, D.C. USA 20005

2. **Stop Clear Cutting-** Work with organizations that work
   to eliminate clear cutting: Search Stop Clear Cutting
   California (list- your particular state) 7

3. **Environmental Investigation Agency-Deforestation**
   search 8

4. **Trees for the Future-** This is a story by Steve Saunders
   who found what one person can do to help reduce CO_2
   in the environment:

"Simply plant trees to act as carbon sinks. After planting a
number of trees where we lived and running out of room, I sought
another way of planting trees. I found *Trees for The Future* who
plant trees all over the globe for a cost of ten cents per tree.
   Search: Trees for the Future 9
   Through donating $3,500. to this group, I've been able to spon-
sor the planting of 35,000 trees. About nine years ago I joined

Facebook and opened a *Trees for The Future* page. We had over 2000 members and managed to raise $1000 in additional donations to Trees. The page I opened was part of a donations group which no longer exists. 10

My story is small but does show what one person can do to help curb increasing CO_2. To effect any meaningful change, many individuals will need to become involved. Perhaps this book will help inspire others to choose career paths which will successfully meet the environmental changes which lay ahead." Interview with Steve Saunders, Big Sur, CA. 2019 11

5 Rainforest Alliance

Search Rainforest Alliance for information 12
This Alliance states that it "protect forests, improve the livelihoods of farmers and forest communities, promote their human rights, and help them mitigate and adapt to the climate crisis."

Concentrating on Guatemala, Southeast Asia, the Amazon region and other deforested places, Rainforest Alliance, focuses on key crops linked to tropical deforestation (such as coffee, tea, cocoa, and bananas); to date, there are more than two million farmers in their agricultural certification program who are practicing more sustainable growing methods. These techniques are designed to maximize the productivity of existing farmland in order to prevent encroachment into the forest, including: boosting soil health through composting, integrated weed management, crop rotation, and climate-smart techniques for conserving water and preventing diseases.

Write, call or email for Inquiry:
125 Broad Street, 9th Floor
NewYork, NY 10004 USA
Phone: +1 (212) 677-1900
Fax: +1 (212) 677-2187
Email: info@ra.org

Search for Rainforest Alliance Careers or Contact: 13

6. Conservation International
Search for Information. 14

Want to get involved with finances to help heal critical areas on the planet? Then hear what this organization has to say: "Conservation International works to find the funding to ensure that the most critical areas on Earth — places that provide us with our food, our water and more — remain intact. We set up trust funds for protected areas. We provide grants to help people maintain their water supplies. And we help finance work that protects regions rich with life — all to secure a future for Earth's most critical life support systems."

Search Conservation International or Contact for Careers: 15

Contact for Inquiry:
2011 Crystal Drive, Suite 600
Arlington, VA 22202
Phone: 1.703.341.2400

7. **Consumer Solution-** For consumers, it makes sense to examine the products and meats you buy, looking for sustainably produced sources when you can. Nonprofit groups such as the Forest Stewardship Council 16 and the Rainforest Alliance 17 certify products they consider sustainable, while the World Wildlife Fund has a palm oil scorecard for consumer brands. They indicate with companies scoring an average of 13.2 points out of 24 and not one single one achieving full marks, the 2021 scorecard is a reminder of the tremendous progress that has yet to be made. But the commendable performance of some companies is proof that sustainable palm oil is achievable.

8. Search: **Pachamama Alliance** and see how you can help:
https://pachamama.org/
Presidio Bldg. #1009

P.O. Box 29191
San Francisco, CA 94129
T +1 415 561 4522

Let's not forget that any of these actions will also help:
- the winged ones
- the crawlers
- the swimmers
- the hoofed and claw-toed ones
- the microscopic ones

Wind Turbines

Chapter 9

Air As
Renewable Energy

The Office of Energy Efficiency and Renewable Energy defines renewable energy as energy produced from sources like the sun and wind that are naturally replenished and do not run out. Renewable energy can be used for electricity generation, space and water heating and cooling, and transportation.

Non-renewable energy, in contrast, comes from finite sources that could get used up, such as fossil fuels like coal and oil. [1]

With air as one of the sources of renewable energy, I can think only of windmills that are now called wind turbines to harness the power of the wind. A wind turbine could be a source of renewable energy for the national grid or for small isolated communities, depending on the size. Wind can be harnessed on a large scale by "wind farms" using high-tech wind turbines. These farms are found throughout the world. There are off-shore wind turbines found in the North Sea and off the coast of Belgium for that matter.

The main concern is that harnessing wind only happens when the wind blows, meaning there must be backup power generation systems in place as well. Other concerns are impact upon scenic landscape and that sea birds may be killed by the large rotors. 2

Even taking this into consideration, wind is still a better option than tearing down mountains, destroying ecosystems or endangering our lungs by the emitted CO_2 into the environment caused by fossil fuels. This powerful source of energy consists of converting the energy produced by the movement of wind turbine blades driven by the wind into electrical energy.

There are benefits: It reduces the use of fossil fuels; it is plentiful; reduces energy imports, creates wealth and local employment, is not a pollutant and is a plus to sustainability.

These are all good reasons to rely on wind energy and yet there is a good reason to rely on the expertise and technology of offshore gas and oil. According to the Centre for Technology, "Offshore wind requires specialized skills, equipment and services. Thanks to technology overlap, notably in the maritime and subsea areas, it is not an insurmountable step for oil and gas companies to adapt and redeploy their existing technologies into offshore wind." 3

To further this fact, additional research indicates that "diversification may enable companies to reduce dependence on a single market and to reorient and distribute resources between markets according to the situation – retreating from or returning to their core market" 4

So, let's see which way the wind blows in the opinion of an Explorer publication about offshore oil and gas for offshore wind. One major player leading the way in wind turbines called Equinor is appearing as the world's leading offshore wind developer. By using its offshore gas and oil knowledge, they pioneered a fully operational floating wind farm, Hywind Scotland, which powers over 22,000 homes in the United Kingdom. 5

ACTION

THE NEW STORY OF HEALING & SOLUTIONS

Every solution and resource births a new story for change. If we want to see changes, we must take action.

In this part of the chapter are possible solutions and resources to help inspire actions so we can have opportunities to live new or revived stories to:

- Awaken to the consequences that wind energy is a powerful source of renewable energy to use instead of fossil fuels

- Find any action item as a pathway towards a climate career, social justice career, research or volunteer project that is just waiting for you to choose from. That is, unless you have already been triggered into action.

Resources to Consider:
In the last seven chapters, please search for the action plan you might pursue to help heal the planet and its people. Additionally, there are wind farms and job resources you might search for under their titles.

Wind Farms
• Equinor Wind Farm 6
Norway based energy company that is pursing renewable energy opportunities on the west and east coast of the US. "Empire Wind" project is planned for 80,000 acres of federal waters south of Long Island which plans to power 1 million NY homes. (potential capacity of 2,000 MW with each turbine to have an installed 10-15MW capacity (very high)

• Alta Wind Energy Centre 7
Second largest onshore wind energy center in the world located in the foothills of the Tehachapi Mountains in Kern County.

600+ wind turbines with 1,550 MW capacity supply power to over 275,000 homes in Southern California

Phone: 646-829-3900
Email: contact@terra-gen.com

Los Vientos Wind Farm 8

Duke Energy Renewables constructed the wind farm in Star County Texas in five phases with an estimated capacity of 2100 MW of renewable energy. Five phases of construction with an estimated capacity of 2100 MW of renewable energy established them in the top ten for wind production in the country.

Get Involved with Manufacturers
Search the following public manufacturers for more information:

Wind Turbine Manufacturers: 9

1) Siemens
   a. Current MW Capacity: 52,770.85

   b. 100+ GW installed capacity with over 90 countries with 29,00+ serviced.
   Email: support.energy@siemens.com
   Jobs: Validation Engineer, HVAC Systems Engineer, Weight Systems Engineer, etc.

2) Vestas
   a. Current MW Capacity: 60,6979.20

   b. They design, manufacture, install, and service wind turbines across the globe, and with more than 100+ GW of wind turbines in 80 countries, Vestas is #2 for installed wind power.

   c. Vestas' employees more than 24,500 around the globe

and currently boasts ≈ 23% of the wind industries market share.

Email: vestas@vestas.com

Jobs: https://careers.vestas.com/

Address: Hedeager 42 8200 Aarhus N, Denmark (+45) 97 30 00 00

3) GE Renewable Energy
   a. Current MW Capacity: 52,949.89

   b. GE Renewable Energy has installed more than 400+ gigawatts capacity globally to make the world work better and cleaner. With more than 22,000 employees present in more than 60 countries, GE Renewable Energy is backed by the resources of the world's first digital industrial company.

4) Enercon:
   a. Current MW Capacity: 17.660.29

   b. Based in Aurich, Lower Saxony, Germany, is the fourth-largest wind turbine manufacturer in the world and has been the market leader in Germany since the mid-nineties.
   c. As of December 2017, Enercon had installed more than 26,300 wind turbines, with a power generating capacity exceeding 43 GW. The most-often installed model is the E-40, which pioneered the gearbox-less design in 1993.
      Contact: Teerhof 59, D-28199 Bremen, Germany
      Phone: +49 421/ 24415100
      Fax: +49 421/2441539
      E-mail: sales.international@enercon.de

5) Nordex SE: 11,708.12 MW
   a. Current MW Capacity:

b. **Nordex SE** has installed wind power capacity of more than 23 GW in over 25 markets. It currently has roughly 5,400 employees. The production network comprises plants in Germany, Spain, Brazil, the United States, and India. The product range primarily concentrates on onshore turbines in the 1.5 – 4.8 MW class addressing the requirements of land constrained as well as grid constrained markets.

c. Langenhorner Chaussee 600 22419 Hamburg, Germany Tel.: +49 - (40) - 300 30 – 1000

Email: info@nordex-online.com

Let's not forget that any of these actions will also help:
- the winged ones
- the crawlers
- the swimmers
- the hoofed and claw-toed ones
- the microscopic ones

The Elemental Mirror of Water

Water

What's in the Water that mirrors our world?

What's in the Water that nourishes or perishes?

What's happening with Water in our overly-heated environment that affects both planet and people?

What goes on with Water-related disasters that is justifiable for struggling people or people-of-color in our world?

Here's what the late West African writer and teacher, Malidoma Patrice Some, once said about water:

"People, especially people in crisis, are naturally attracted to water. Just the sight of a large body of water brings a feeling of quiet and peace, a feeling of home. Water resets a system gone by. In this western culture, the most crucial task requiring the reconciling energy of water may be the confrontation of overwhelming contradictory emotions carried by people."

Even if we don't know all the answers right now, let's open

up to some of these issues and solutions presented in the next four chapters of this section. Please commit to peacefully act on at least one or some of them to help create a better world!

Thus far, what are your thoughts to help create a better world?

Washed out neighborhood road in Big Sur.

CHAPTER 10

Local Storms And Flooding

In writing this chapter about storms and floods, I had firsthand experience with a catastrophic rainstorm in February of 2017 in Big Sur, CA. This affected a vast area of the Big Sur coastline. No one could drive in or out from the North or the South.

This storm created havoc in this coastal area causing major mud slides, namely: Mud Creek Slide, Paul's Slide; and a slowly sliding Pfeiffer Bridge with loosened support beams holding on by threads of mud rather than fixed soil. The intense rainfalls not only disintegrated this major bridge, but also caused flooding in Andrew Molera State Park. Some watersheds flowing from the Santa Lucia Mountains along the Big Sur coast flooded the main road, Highway 1. The storm caused private roads to fall to pieces, disrupting internal travel in my neighborhood. These occurrences divided the Big Sur community for about 8 months. The old and yet healthy method of getting from one place to another is called walking. With the community teaming with the State Park and other agencies, a walking trails against the mountainside was made. It slowed the pace which I thought was a good thing that came out of the catastrophe. Cars were parked at the trailhead as many locals packed their groceries and supplies on their backs to get to their homes.

It created an economic impact upon those who depended on the tourist industry for businesses and jobs. This caused many folks to re-locate. Although Big Sur has experienced many storms, this particular change in the climate resulted from what is called an atmospheric river.

Described by National Oceanic & Atmospheric Administration (NOAA), atmospheric rivers are relatively long, narrow regions in the atmosphere – like rivers in the sky – that transport most of the water vapor outside of the tropics. These columns of vapor move with the weather, carrying an amount of water vapor roughly equivalent to the average flow of water at the mouth of the Mississippi River. When the atmospheric rivers make landfall, they often release this water vapor in the form of rain or snow. 1

Although atmospheric rivers come in many shapes and sizes, those that contain the largest amounts of water vapor and the strongest winds can create extreme rainfall and floods, often by stalling over watersheds vulnerable to flooding. These events can disrupt travel, induce mudslides and cause catastrophic damage to life and property. A perfect example that resulted from an atmospheric river that occurred on the Big Sur coast is the Mud Creek slide.

Mud Creek slide

Paul's slide also divided the Big Sur coast, but did not cause the major havoc and reconstruction of the Highway 1 like the Mud Creek slide that plummeted parts of Highway 1 into the ocean.

Not all atmospheric rivers cause damage; most are weak systems that often provide beneficial rain or snow that is crucial to the water supply. Atmospheric rivers are closely tied to both water supply and flood risks — particularly in the western United States. [1]

Taking a step off the contemporary path, let's take a look at a pathway to deal with flooding by mimicking nature. What if another atmospheric river plagues Big Sur or any area and creates an enormous amount of flooding? How can mimicking nature help this situation? What is it all about?

This branch of science called Biomimicry was created by Janine Benyus, President of the Biomimicry Institute, that "bridges nature and design." That leads us with a lot of possibilities to explore. When Dr. Benyus spoke at the Bioneers Conference in northern California in 2016, [2] I was struck with amazement and awe at the complexity yet simplicity of what nature has to offer. It is further described by the biomimicry organization as "learning from and then emulating natural forms, processes, and ecosystems to create more sustainable designs. It's studying a leaf to invent a better solar cell or a coral reef to make a resilient company. The core idea is that nature has already solved many of the problems

137

we are grappling with: energy, food production, climate control, benign chemistry, transportation, collaboration, and more." Dr. Benyus has indicated that the more our world functions like the natural world, the more likely we are to endure on this home that is ours, but not ours alone. 3

Some biomimicry examples:

1. The Wright brothers copying a vulture's wings to invent the first airplane

2. Gardeners mimic forest floors by keeping soil undisturbed; that's creating continuing composting that equires less water as well as attention

3. Alexander Graham Bell's telephone speaker and receiver idea came from understanding how the human tongue and your drum works

4. Alaskan hunters mimicking how the polar bear stalk seals

Further research Biomimicry https://biomimicry.org/what-is-biomimicry/ or https://asknature.org/

What can one do or ameliorate (restructure or improve) an area in event of a global warming disaster where floods occur? As I ask this question, I am constantly thinking of what happens to cities and to people when there are storms, tsunamis, and heavy weather conditions that create this. In other words, how can nature help us to understand how to collect, move and absorb water?

Rather than have cities, businesses, villages, and homes destroyed, not to mention totaled vehicles, is there a way to pump out, absorb, dry and replenish without outbreak of diseases? What if things like this could prevent flooding catastrophes?

Let's dig into this matter a bit deeper with more scientific information.

According to the Ocean Institute, "The oceans cover 71 percent of the Earth's surface and contain 97 percent of the Earth's water. Less than 1 percent of the Earth's water is fresh water, and 2-3 percent is contained in glaciers and ice caps. The oceans contain 99 percent of the living space on the planet." 4

With that said, let's continue this chapter with this large body of water. From the CO_2 concentrations in the atmosphere, we'll go down into the ocean where it is said that the ocean temperature is rising.

From an oceanographic perspective, I was curious about climate change several years ago. I conducted an interview with Dr. George Somero, David and Lucille Packard Professor in Marine Science Emeritus who currently works at Stanford University's Hopkins Marine Station in Pacific Grove, CA. 5 He is also referenced in the opening chapter of this book regarding IPCC. He clearly stated, "Go back a century when the arctic had a lot of more ice and snow that was reflecting sunlight back into space but now with less ice on the ocean, the water absorbs most sunlight.

If you have greater brown earth instead of white covered earth, you are just absorbing more energy, that warms the earth. When you absorb more sunlight that will have an effect on the atmosphere. This will also affect for example, the position of the jet stream (fast flowing, narrow air currents found in the atmosphere). To elaborate here, jet streams are relatively narrow bands of strong wind in the upper levels of the atmosphere. The winds blow from west to east in jet streams but the flow often shifts to the north and south. Jet streams follow the boundaries between hot and cold air. The earth's rotation is responsible for the jet stream as well. 6

So, one thought might be the weather we had just up until two weeks ago (before the interview). It was very dry and the jet stream was carrying storms into the Midwest and the East Coast. They were getting hammered while we (in California) we were drying out. Normally the jet stream would bring some storms down to us which it is finally doing but the thought is that when you change the amount of heat up in the Arctic, it is going to affect the atmosphere. It is just a hypothesis but it is still one way in which warming could lead to more severe electric storms in some parts of the country."

The question to Dr. Somero: **Are oceans actually warming?** His answer was an affirmative, "yes". My next question was, how do we measure the heat?

"Well you can measure that as we do here at the Marine station. We (Stanford University's Marine Hopkins Lab) have been testing the temperature of the ocean since 1920. It has heated by a degree Celsius and that is kind of true all over the ocean and it doesn't sound like much."

The next question: **Does the warming ocean have its harmful effects with air flow and jet streams?** "It can, and in some parts of the ocean warming is occurring a lot faster at higher latitudes. It is occurring much faster than it is in the tropics. And the reason is about atmospheric flow. So, when you talk about a 1-degree change, over the last century, you know there are some areas that have gotten colder and there are other areas that have gotten a lot warmer and that is about the average and the Polar Regions have seen generally about the highest rate of temperature increase.

As Somero states that warming could lead to more severe rain storms, there are scores of unusual storm weather patterns. For instance, Hurricane Florence in 2018 that made landfall along the southeastern coast of North Carolina as a Category 1 storm. It also brought significant storm surge flooding to portions of eastern North Carolina. Also, in 2018, Hurricane Michael was the strongest hurricane to ever hit the Florida Panhandle. Looking back to 2017, six hurricanes became major storms in 2017-Harvey, Irma, Jose, Lee, Maria, and Ophelia. Irma was one of the most powerful and costliest hurricanes in the Atlantic Basin. And we can't forget Hurricane Katrina back in 2005 as the strongest to make landfall in the US as it compares to Hurricane Wilma. **7**

In Puerto Rico, Hurricane Maria took a big toll on not only land but on lives with almost 3,000 deaths effected by the big hit according to a study conducted by George Washington University. 8

The term "rapid intensification" or highly intense winds is generally measured by comparing the strength of a hurricane over a 24-hour period. A change in storm wind speed of greater than 35mph in 24 hours is generally the cutoff. The destructive storms mentioned went through rapid intensification. 9

As with most storms, it's not just the winds that do the damage, it's also damage and contamination of the water. Overflows with ocean water can impact farmlands, homes, water systems and most everything that water can travel to. An example would be the coral island nation of Tuvalu in the Pacific.

Let's turn again to greenhouse gases and its effects that may be the primary factor creating intense weather patterns. 10 Dow & Downing refer to the greenhouse gas effect as a "Heat-distribution Engine." This engine relies on atmospheric and oceanic circulation also known as jet streams to move heat energy and distribute it more evenly around the world. Dr. George Somero has described previously about this as well. 11

Further, Dow & Downing points out that the equator is where most heating occurs. All year round the sun's rays are almost perpendicular [vertical or upright]. Both the circulation from the atmosphere and the ocean contribute equally in moving energy from the equator towards the poles. It is the heat energy distribution that creates our climates, so when the global systems change, the climate changes. The results could be very hot summers, very little snowfall and rainfall with a frequency and intensity of storm outcomes.

Dow and Downing state outright, that "increased temperatures and water vapor in the air over tropical oceans creates conditions for cyclones, hurricanes and typhoons." And that warming waters in the Pacific triggers El Nino events (severe weather conditions that do not usually take place in a certain region). They note that during El Nino years, rainfall follows the warm water that has led to flooding in Peru, drought in Indonesia and Australia and disruptive weather patterns throughout the world. 12

ACTION

THE NEW STORY OF HEALING & SOLUTIONS

Every solution and resource births a new story for change. If we want to see changes, we must take action.

In this part of the chapter are possible solutions and resources to help inspire actions so we can have opportunities to live new or revived stories to:

A) Awaken to the consequences of unexpected weather patterns that brings forth hurricanes and storms that leave little time to act.

B) Find ways to help towns, villages and cities be prepared to help the vulnerable populations to be immediately rescued from floods and storms with appropriate emergency housing that lends itself to some degree of privacy.

C) Find any action item as a pathway towards a climate career, social justice career, research or volunteer project that is just waiting for you to choose from. That is, unless you have already been triggered into action.

Possible Solutions to Help Flood-Prone Areas

Big Sur biologist, Jessica Koning, researched some biomimicry solutions and she was surprised at her findings. "I thought I would be looking exclusively into how plants and animals in flood-prone areas protected themselves. However, the best source found was far more creative and cleverer than that."

A. The Elephant Foot Plant lives in an extremely drought-prone part of Africa. Its base is wide and looks similar to an elephant's foot. When it does rain, it is very important for this plant to be able to store as much of

142

the rainfall as possible. This plant has a sponge-like above ground root that is capable of expanding many times over to store rainwater. The study suggested that putting something like a warehouse with open windows half-full of dry sponges at flood stage could contain excess water. So interestingly, an adaptation to deal with drought can be reapplied to floods! 13

B. European water voles live along stream banks. They look similar to a rat, yet they love being around water. Their burrows are designed NOT to flood if the stream gets high. The entrance hole is lower than the living chamber, and if this isn't enough, there are unused chambers that are supposed to flood so the living chambers DON'T. The study designed floodwater storage areas by placing tunnels along stream beds that were mirror images of the water voles' home. So, by building a tunnel at the OPPOSITE angles of what the voles does, you create an effective water catchment area. 14

C. The Giraffe has a complicated system of valves (one-way gates) in the blood vessels of its neck, the long neck that allows the giraffe to eat vegetation inaccessible to other terrestrial creatures can create extreme changes in blood pressure in the brain as the giraffe's head goes up and down. This could cause the giraffe to faint or feel light-headed if it raises its head too swiftly (low blood pressure) or have a stroke (brain bleed) if it lowers its head quickly (high blood pressure), These valves have the job of adjusting the blood volume to keep the giraffes' blood pressure in its brain constant. A similar system of valves and pumps could be programmed to combat floods using the same feedback systems the giraffe uses when lowering its head swiftly. A flood would be analogous to high-blood pressure, the giraffes' heart would be the water pumps, the blood vessels would be the water tanks, and the valves would be the leakproof that would seal the water tanks! 15

D. Koning goes on to say that if she "were to look for a biomimicry system that might work in a rural area like Big Sur, CA and attempt a lower impact on the environment, then she would propose the California Coastal Redwood as a biomimicry model. Redwoods typically grow along streams. They require quite a bit of water, and living along a river satisfies this need very well." She goes on to say that a tree can't evacuate. When there is a flood, they are stuck there the same way we can't move a house with a foundation (although we can evacuate ourselves and possessions if we have time). The main danger of the redwood is not the high water, but the silt (fine dirt particles) that the flood water carries. This silt settles on the ground, and makes the soil deeper. Redwoods have shallow roots that need air. If those roots are deeply buried, the roots suffocate and the tree dies.

Redwoods survive floods by abandoning those buried roots and constructing a new root system closer to the surface. They essentially build a new foundation on top of the old one. Therefore, the redwood models suggest that permanent foundations are not feasible long-term solutions for structures immediately adjacent to rivers. If a home doesn't have a permanent foundation, then the floods come, you can either move it (like a trailer home or yurt), or design your house in a modular stackable way so it could become your new foundation if it were badly flooded. You would bolt your new house on top of your old one if a flood filled it with water, rocks and /or silt. Obviously, it would require the previous house to be leveled, filled with concrete, and sealed to provide a sturdy base for your "new" house. 16

Other Possible Solutions to Help
Flood Prone Areas

Floating Cities
Research: the LEGO House and other architectural forms of the future" by Bjarke Ingels, can give some interesting ideas. 17

Further research: Biomimicry 18

Mangrove Bushes
Another naturalistic thing we can do to protect warm region coastal areas from flooding is to use a natural barrier. Plant mangrove bushes and other coastal barrier plants. These are found in Florida. They create barriers to storm waves in the form of shoals, salt marshes. This is great for developing nations who cannot afford to build coastal defenses. Check out the projects below:

Here are some additional Resources that you can explore to help with Flooding and Storm Solutions:

• **Mangrove Action Project** 19
Extensive network of forest communities, grassroots NGO's, researchers, and local government to provide training sessions in 15 countries around the world on hand planting methods for mangrove restoration. 250,000+ students taught worldwide and restoration techniques.
1 (206)-207-2022, 1001 4th Ave, #3200 Seattle, WA 98154

• **Global Mangrove Alliance** 20
Mangrove knowledge hub that serves as a clearing house for information that is widely accessible on their website. Various initiatives that incorporate NGO's local governments and local stakeholders including: Creating a Blue Carbon working group to increase recognition and application of blue carbon ecosystem values to enhance management and support for coastal wetlands, training and education for beach forest tree propagation, and land use planning for identifying and protecting mangroves in the face

of development.

• **Tetra Tech** 21

Tetra Tech helped pioneer the concept of green infrastructure during the 1990s and has continued to provide cutting-edge approaches as the focus of storm water management has evolved to meet regulatory and environmental goals. In January 2017, *Climate Change Business Journal (CCBJ)* recognized Tetra Tech's green infrastructure work at Viola Liuzzo Park in Detroit, Michigan, as among the most outstanding projects of 2016. In January 2017, *Climate Change Business Journal (CCBJ)* recognized Tetra Tech's green infrastructure work at Viola Liuzzo Park in Detroit, Michigan, as among the most outstanding projects of 2016.

Charlie MacPherson
Media and Public Relations
3475 East Foothill Boulevard
Pasadena, California 91107-6024
USA
+1 (626) 470-2439
charlie.macpherson@tetratech.com

Let's not forget that any of these actions will also help:
- the winged ones
- the crawlers
- the swimmers
- the hoofed and claw-toed ones
- the microscopic ones

Coral Bleaching

Chapter 11

Coral Reef Disruptions

First of all, let's get a simple Wikipedia definition of coral reefs: "A coral reef is an underwater ecosystem characterized by reef-building corals. Reefs are formed of colonies of coral polyps held together by calcium carbonate. Most coral reefs are built from stony corals, whose polyps cluster in groups.

Coral belongs to the class Anthozoa in the animal phylum Cnidaria, which includes sea anemonesand jellyfish. Unlike sea anemones, corals secrete hard carbonate exoskeletons that support and protect the coral. Most reefs grow best in warm, shallow, clear, sunny and agitated water. 1

Basically speaking, there are four types of reefs that are found in various parts of the world that reflect injustice on marine life and human life. These types of reefs according to Different Type network, include:

Fringing Reefs- Fringing reefs grow very close to the shoreline. However, they did not originate from the shoreline. This type of reef is most common in the Bahamas. There are also plenty of fringing reefs in the Red Sea. Many types of fish make their home within a coral reef.

Barrier Reefs- the barrier type reef is separated from the shoreline by either a deep channel or a lagoon. The most famous barrier reef system in the world can be found off the coast of Australia. It is aptly named 'the Great Barrier Reef'. The Great Barrier Reef has almost 3,000 individual reefs and over nine hundred islands within its ecosystem that occupies 344,400 square kilometers. It is completely made by living organisms smaller than your finger. Other notable barrier reef systems in the seas are the New Caledonian Barrier Reef, the Mesoamerican Barrier Reef, and the Belize Barrier Reef systems.

Atoll Reefs- An atoll reef looks like a patch of land surrounding an inner lagoon from the air. The formation of an atoll is a very complicated process. Coral reefs form around an island. Over time, so much time, the island sinks under the sea. As the island begins sinking, the gap between the new shores and the reef is filled by new coral. This process repeats itself until a mountain tip is what remains of the island. The corals build their colonies around that tip. Finally, even the tip of the island sinks. The corals, who need light, build their colonies where they can get the most light. In case of a volcanic island, this is the rim of the caldera. As time passes again, the island continues to sink lower, and the corals keep building upward. Eventually the reef that was built on the rim is the only visible trace of the coral reef. The end result is an atoll.

Patch Reefs- A patch reef is a minor type of reef. It is so named because it is usually an isolated patch of small, coral reef growth. In the Florida Keys, patch reefs are a very common occurrence. Between Soldier Key and North Key Largo, there are more than four thousand identified patch reefs. Just like any coral reef, patch reefs need light to survive, and grow very close to shore. 2

Coral reefs provide protection and shelter for many different species of fish. Without these coral reefs (ridges of coral just above the surface of the sea) fish are left homeless with nowhere to live and nowhere to have their babies. This could cause mass extinction for some marine life when water is acidified causing a disruption or disturbance. This happens when waters absorb carbon dioxide, making it nearly impossible for organisms such as crabs, clams and microscopic plankton to build shells, thus destroying their food supply.

To clarify scientifically, National Oceanic and Atmospheric Administration (NOAA) has noted the following:

When carbon dioxide (CO_2) is absorbed by seawater, chemical reactions occur that reduce seawater pH, carbonate ion concentration, and saturation states of biologically important calcium carbonate minerals. (*A pH unit is a measure of acidity ranging from 0-14. The lower the value, the higher the acidity of the environment. A shift in pH to a lower value reflects an increase in acidity*). These chemical reactions are termed "ocean acidification," recognized here as:

1. Continued ocean acidification that is causing many parts of the ocean to become undersaturated with these minerals, which is likely to affect the ability of some organisms to produce and maintain their shells.

2. The pH of surface ocean waters has fallen by 0.1 pH units since the beginning of the Industrial Revolution. This change represents approximately a 30 percent increase in acidity, like the Richter scale, the pH scale is logarithmic.

3. By the end of this century the surface waters of the ocean could have acidity levels nearly 150 percent higher, resulting in a pH that the oceans haven't experienced for more than 20 million years.

4. Further studies from NOAA have shown that lower environmental calcium carbonate saturation states can have a dramatic effect on some calcifying species, ncluding oysters, clams, sea urchins, shallow water corals, deep sea corals, and calcareous plankton. Today, more than a billion people worldwide rely on food from the ocean as their primary source of protein. Thus, both jobs and food security in the U.S. and around the world depend on the fish and shellfish in our oceans.

5. Increasing ocean acidification has been shown to significantly reduce the ability of reef-building corals to produce their skeletons. 3

In a study published in Proceedings of the National Academy of Sciences of the USA, coral biologists reported that ocean acidification could compromise the successful fertilization, larval settlement and survivorship of Elkhorn coral, an endangered species. These research results suggest that ocean acidification could severely impact the ability of coral reefs to recover from disturbance. Other research indicates that, by the end of this century, coral reefs may erode faster than they can be rebuilt. 4

At the same time, there are other incidences that disrupt coral reef. According to the Environmental Protection Agency (EPA), here are some local threats:

• **Physical damage** or **destruction** from coastal development, dredging, quarrying, destructive fishing practices and gear, boat anchors and groundings, and recreational misuse (touching or removing corals).

• **Pollution** that originates on land but finds its way into coastal waters. There are many types and sources of pollution from land-based activities, for example:

A. **Sedimentation** from coastal development, urban storm water runoff, forestry, and agriculture nutrients (phosphorus and nitrogen) from fertilizer runoff and sewer discharges.

Sedimentation has been identified as a primary stressor for the existence and recovery of coral species and their habitats. Sediment deposited onto reefs can smother corals and interfere with their ability to feed, grow, and reproduce.

Nutrients are generally recognized as beneficial for marine ecosystems; however, coral reefs are adapted to low nutrient levels; an excess of nutrients can lead to the growth of algae that blocks sunlight and consumes oxygen that corals need for respiration. This often results in an imbalance affecting the entire ecosystem. Excess nutrients can also support growth of microorganisms, like bacteria and fungi, that can be pathogenic to corals.

B. **Pathogens** from inadequately treated sewage, storm water, and runoff from livestock pens.

   Although rare, bacteria and parasites from fecal contamination can cause disease in corals, especially if they are stressed by other environmental conditions. Coral disease occurs in healthy ecosystems, but the input of pathogen-containing pollution can exacerbate the frequency and intensity of disease outbreaks.

C. **Toxic substances**, including metals, organic chemicals and pesticides found in industrial discharges, sun screens, urban and agricultural runoff, mining activities, and runoff from landfills.

   Pesticides can affect coral reproduction, growth, and other physiological processes. Herbicides, in particular, can affect the symbiotic algae (plants). This can damage their partnership with coral and result in bleaching. Metals, such as mercury and lead, and organic chemicals, such as polychlorobiphenyls (PCBs), oxybenzone and dioxin, are suspected of affecting coral reproduction, growth rate, feeding, and defensive responses.

D. **Trash and micro-plastics** from improper disposal and storm water runoff. Trash such as plastic bags, bottles, and discarded fishing gear (also called marine debris) that makes its way into the sea can snag on corals and block the sunlight needed or entangle and kill reef organisms and break or damage corals. Degraded plastics and microplastics (e.g., beads in soap) can be consumed by coral, fish, sea turtles, and other reef animals, blocking their digestive tracts and potentially introducing toxins.

• Overfishing can alter food-web structure and cause cascading effects, such as reducing the numbers of grazing fish that keep corals clean of algal overgrowth. Blast fishing (i.e., using explosives to kill fish) can cause physical damage to corals as well. Trawling nets also have the ability to destroy or break apart structural coral reefs.

• Coral harvesting for the aquarium trade, jewelry, and curios can lead to over-harvesting of specific species, destruction of reef habitat, and reduced biodiversity. 5

Nonetheless, coral reef bleaching socially impacts sea creatures as has been explained. Beyond this catastrophic injustice, climatically and man-caused, human lives are affected. According to NOAA, approximately 50 million people worldwide depend upon reefs for food and their livelihoods. 6

In areas such as the Great Barrier Reef, human survival and medical research can be lost with coral reef disruption. 7 According to Ukessays, a dissertation and information service, The Great Barrier Reef provides people with valuable natural resources, such as foods and drugs. Corals provide housing and shelters, they feed fish, and produce the other marine animals that humans eat. According to *Saltwater Science* published in *Nature* by Jessica Carilli, "some estimates say that over 1 billion people depend on food from coral reefs" (Carilli, 2013). Without the Great Barrier Reef, people will be forced to find new sources of food. The prosperous environment of coral reefs has also helped scientists and researchers discover new medical advancements. According to the *The Nature Conservancy*, "the study of marine life has enabled the development of treatments for diseases like asthma, soft-tissue sarcoma, and certain lung cancers. Scientists are also reportedly using bivalves like clams from the Great Barrier Reef to study the aging process, and molecules in certain aquatic invertebrates could have anti-viral, antibacterial, and anti-tumor properties" (The Nature Conservatory, 2018). As the coral in the Great Barrier Reef becomes bleached, the resources people are harvesting from the reef are diminishing and becoming unavailable. The more coral that dies, the less people have, which will impact the lives of millions of people as they have to find new resources to consume.

In addition to providing food and medicine, the Great Barrier Reef also provides protection to the tourist-reliant coastlines. According to "The terrible things that would happen if all the coral reefs died off" published in *Business Insider*, "the reefs act as natural barriers, canceling out 97% of a wave's strength and protecting more than 200 million people…. Building seawalls for the same protection costs $2.5 million per mile" (Sharma & Reilly, 2018). Without the Great Barrier Reef, the town of Queensland, Australia will become more susceptible to natural disasters since there will no longer be any coral to protect the town, potentially harming thousands of people. The Great Barrier Reef also acts as a place for social interactions among people. Recreational activities, such as those found in the Great Barrier Reef, create cultural and social experiences for people like to visiting and that can only be found in the World Heritage spot. 8

In contrast, let's reflect in the mirror at the crossroads on other social justice concerns when people of color were compelled to whiten or bleach out their skin. This could be viewed as a metaphor and yet it's true. So, what dies off? Could it be assimilation into a predominately white culture in order to be mostly accepted and treated better? Does this mean beauty, for those who bleach their face or is it a form of de-valuing one's own cultural heritage?

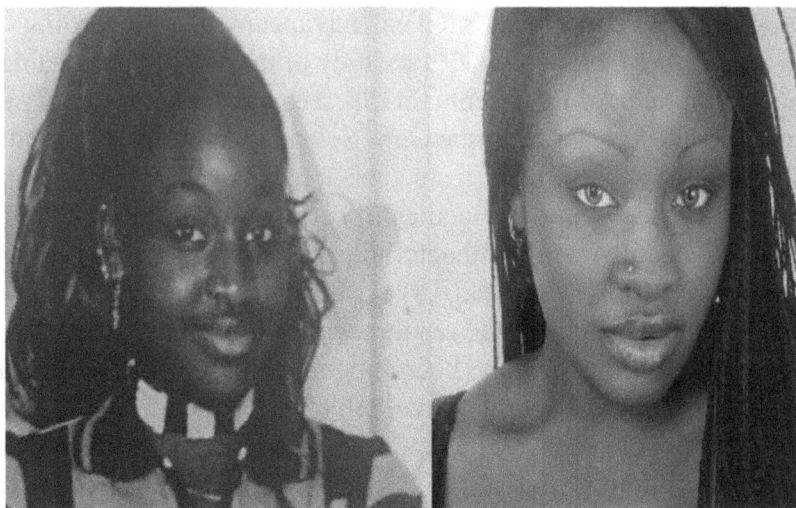

Bleach My Skin

Skin bleaching is done on a small scale today and was once very common to use this toxic lye-containing crème. The physical applying to one's face can be thought of as psychological harm placed on people of color in order to make their facial skin look lighter. And of course, underneath the surface, came the straightening of kinky hair.

Once, the light-colored skin and the straight hair folks thought, they would have a better chance of obtaining privileges and favor, and is somewhat true even today. Lately, Black folks have come to terms, often legally, in order to be able to wear their hair and hair accessories in a style more fitting to them as defined by their heritage.

As long as we are getting really personal with the face and hair, let's mirror what is reflected with black swimmers, especially those in athletic situations who are harassed for being themselves with their natural hair types. See if you can figure out what is happening beneath the surface in this human situation. The International Swimming Federation rejected the use of Soul Cap, designed for Black Swimmers in the 2021 Olympics, saying it does not "fit the natural form of the head." 9

This statement by the Swimming Federation was made when

Alice Dearing, an African American Woman, qualified for the prestigious Olympics.

"The Federation also said "best knowledge the athletes competing at the international events never used, neither require … caps of such size and configuration."

Considering what is so and accepted by the Swimming Federation, Danielle Obe, the founding member of the Black Swimming Association, told the Guardian the ruling underlined the inherent systemic and institutional inequalities around the sport. For example:

"The original swimming cap, designed by Speedo 50, was created to prevent Caucasian hair from flowing into the face when swimming." Obe said the caps did not work for afro hair, which "grows up and defies gravity." 10

According to the sport's governing body, Swim England, only 2% of regular swimmers are black. It found 95% of black adults and 80% of black children in England do not swim. While 79% of Asian adults and 79% of Asian children do not swim, black children are three times more likely to drown than white children.

Yes, is the answer that hair is an issue as Blacks would rather not have chlorine and water-soaked hair when swimming in order to maintain a presentable look after swimming. This is the reason many Black females avoid swimming. There are just not any swim cap products that really work. And the one that does it, is not found in sport gym stores nor sporting goods stores. Not even the Soul Cap that was designed for Black Swimmers are carried in these stores. Have you figured out what is happening beneath the surface in this human situation? 11

As we look forward, one year later, the International Swimming Federation changed its stance after a period of review and discussions on cap designs. The statement came out Executive Director, Brent Nowicki: "Promoting diversity and inclusivity is at the heart of their [Federation] work, and it is important that all aquatic athletes have access to the appropriate swimwear."

Heaven forbids African Americans who try to change their voice to sound white on the phone in order to be listened to. So, 'you can't expose yourself for who you really are or you will be devoured by predators' is much like the greed of mankind that caused climate warming that affects coral reefs and their inhabitants to be exposed and helpless. The only difference is that the ocean critters can't assimilate to be something else in their old habitats.

ACTION

THE NEW STORY OF HEALING & SOLUTIONS

Every solution and resource births a new story for change. If we want to see changes, we must take action.

In this part of the chapter are possible solutions and resources to help inspire actions so we can have opportunities to live new or revived stories to:

A. Be aware of what lies beneath the surface caused by climate change and by man's actions- be it coral reef disruptions or human intentional bleaching

B. Find any action item as a pathway towards a climate career, social justice career, research or volunteer project that is just waiting for you to choose from. That is, unless you have already been triggered into action.

List of possible action items to pursue:

As concerns social justice issues with what lies underneath the surface that is disrupting, here are some suggested solutions:

- Develop Cultural Awareness. Search for Cultural Competency and Workplace Diversity Trainings or contact https://www.lavernemcleod.com for suggestions.

- Learn to appreciate and value diverse viewpoints of people-of-color

- Resist stereotyping

- Attend Cultural Competency & Diversity Workshops

- Contact your local community colleges and take a course on Diversity & Inclusion.

- Learn to respect and value other cultures and nationalities including your own.

Resources & Solutions to Coral Reef Disruptions:

These are some things that need to happen to help restore coral reefs:

- Creation of coral nurseries to rehabilitate collected broken corals and create new individual colonies

- Construction of artificial structures to provide a growth platform for corals and other reef organisms

- "Plantation" of baby corals on the structures

- Advanced reef restoration technologies (mineral accretion by electrolysis) improving growth and survival rates and increasing the resilience to climate change and other pressures

- Development of coral culture tables for coral aquaculture

Please note that when coral reef mitigations are in effect, many people who rely upon the reefs for survival can be helped as well.

Feel free to explore how you can get involved in the above needed actions by contacting various organizations. Here are some suggestions that you can search:

1. Nature Seychelles

"Nature Seychelles is a leading environmental organization in the Western Indian Ocean. It is the largest and oldest environment NGO in the Seychelles archipelago, where it is involved in environmental conservation and management.

It is an association registered in the Seychelles with a board of Trustees, and a local and international membership since 1998. It is also the **BirdLife Partner in Seychelles**, a member of the **World Conservation Union (IUCN)**, and the Western Indian Ocean Marine Science Association (WIOMSA). 12

Nature Seychelles manages the **world-famous Cousin Island Special Reserve**, one of the oldest marine protected areas in Seychelles.

Nature Seychelles employs 15 full time staff including biologists, an economist, educators and protected area specialists.

The primary objective of Nature Seychelles according to its statutes is to improve the conservation of biodiversity through scientific, management, educational and training programs.

"Reef Rescuers' project was developed to restore the fringing coral reef within Cousin Island Special Reserve. The first-ever large-scale reef restoration project began in 2010 with the financial support of the United States Agency for International Development (USAID). Further financial support was received under the Government of Seychelles-Global Environment Facility (GEF)-United Nations Development Project (UNDP) Protected Area Project in 2011. Utilizing the 'coral gardening' concept, fragments of healthy coral were collected, raised in underwater nurseries and then transplanted onto a degraded reef. Since 2010, 40,000 corals have been raised in underwater nurseries, of which over 24,000 were successfully transplanted, covering the area of a football field (5,225 m2). The long-term success of the project is currently being assessed, with initial data demonstrating an increase in both coral recruit and fish densities following the intervention, highlighting the benefits of active reef restoration."

Contact:
Center for Environment and Education
Roche, Caiman,
P.O. Box 1310, Mahe, Seychelles
+248 460-1100
nature@seychelles.net 13

2. Secore International 14

Leading Conservation Organization for the restoration and protecting of coral reefs. They combine research, reef restoration, education and outreach for their conservation projects. They have reef restoration projects in: The US Virgin Islands, Bahamas, Mexico, Guam and Curacao.

The key aspects to their projects are:

- Producing millions of genetically unique coral larvae by collecting gametes during natural spawning event, followed by in-vitro fertilization. Corals are placed back on the reef. This increases genetic diversity and resilience of natural populations.

- Create tree nurseries and asexual fragments to produce brood stocks for highly endangered coral reefs species which can no longer naturally fertilize.

- Working to create a new approach where coral larvae are settled on specifically- designed substrates. These may be seeded without the need of manual attachment of the coral to structures (which is the most time consuming process)

- Research opportunities to enhance settlement of and survival of coral babies for example, by growing them together with natural grazers such as diadem sea urchins found in the Caribbean.

Contact:
Secore International Rent:
4673 Northwest Parkway Inc
Hillard, OH 43026
614-969-3150
Media Inquiries: sloeschke@secore.org

3. Pur Project 15

PUR Project works with companies to regenerate the ecosystems they depend upon.

They empower local communities to operate long-term socio-environmental projects, we help companies strengthen their supply chains through agroforestry, land restoration and sustainable agricultural practices.

This approach, called "Insetting", encourages our Partners to integrate social and environmental innovation at the heart of their operations. Their main actions are:

1. **Develop:** a global network promoting Insetting solutions from farm to plate.

2. **Promote:** investment in integrated social and environmental projects, leveraging our project, experience and network

3. **Educate:** on why supporting ecosystems is the best possible investment for current and future generations

Pejarkakan Projects: Holistic approach to community-led ecosystem management by tackling marine, coastal and terrestrial environmental degradation. The project combines coral reef restoration, agroforestry, mangrove replantation and a plastic recycling scheme with the aim to empower the local communities to sustainably manage their natural resources.

Coral reef restoration: Over 15 different coral species are planted on the artificial reef. The corals are collected in proximity to the project site and transported the short distance by boat. Only broken coral fragments are collected to avoid damage to the natural reef.

Mangrove replantation: 6 different species are grown in local nurseries before being planted in the field.

Agroforestry: farmers plant a minimum of 3 different species on their farms of a total of 8 species including guava, orange, maoni, gemelina and coconut.

The coral reefs of Pejarakan were destroyed in the last 60 years due to destructive fishing practices such as dynamite and cyanide fishing leading to important losses of marine biodiversity. PUR Project teamed up with the community group Pokmasta to restore the lost biodiversity through the construction of an artificial reef. The artificial structures provide a solid substratum for corals to grow on and will attract a wide range of other marine organisms. To support and sustain coral growth the project uses a technology called Biorock. Biorock consists in passing a low voltage current through the structures – the resulting electrolysis of seawater causes minerals that are naturally present in the seawater to precipitate onto the structures forming a solid limestone coating. Corals growing on Biorock structures are said to be more resilient to pollution and climate change. PUR is also involved in the following:

PLASTIC WASTE MANAGEMENT

Indonesia is the second largest contributor of marine plastic after China; plastic pollution is threatening marine life and ecosystems causing an estimated 13 billion USD in damage every year. The project combines beach cleans with educational events, workshops and trainings, raising awareness for marine pollution as well as coral restoration and preservation. It also includes the implementation of metallic recycling bins and payable recycle collection for shops and businesses.

REFORESTATION & AGROFORESTRY

During the rainy season, heavy rains wash away large quantities of sediments, which are directly discharged into the sea. Sedimentation has a significant negative effect on coral health, hindering restoration effects by inducing additional stress on marine organisms. The project includes the planting of timber and fruit trees around fields, village temples and rivers beds in order to improve soil retention, reduce soil erosion and runoff, enhance water quality within the lagoon for the direct positive impact on coral health, and provide an alternative income source for villagers (fruit and other tree by-product to be sold on local markets).

MANGROVE CONSERVATION & RESTORATION

The Pejarakan project aims also at conserving and restoring mangrove forests, very important for coral health, biodiversity and climate change adaptation (protect coastal areas against storms and floods). In partnership with the Forum Konservasi Putri Menjangan, PUR Project establishes local mangrove nurseries for plantation, restores disused salt farms to productive mangrove systems and pushes for legislative protection.

Contact:
4 rue de la Pierre Levée 75011 Paris, France
PHONE
+33 1 55 28 98 05
EMAIL
contact@purprojet.com

4. Great Barrier Reef Foundation: 16

Over 1 billion people depend on coral reefs for food. The Great Barrier reef has experienced a 50% decline over the past 30 years. Over 70,000 jobs rely on the reef.

"The Foundation was established in 1999 following the first mass coral bleaching of the Great Barrier Reef in 1998, and in alignment with the United Nations World Heritage Convention encouraging countries with world heritage sites to establish a national foundation with the purpose of inviting donations for their protection.

They lead the collaboration of business, science, government and philanthropy – groups who would not otherwise come together – for the benefit of the Reef. Our success is due to the quality of institutions and people we bring together – harnessing advances in science, technology and industry to ensure a future for this global treasure."

"Working closely with the Great Barrier Reef Marine Park Authority, they fund priority projects that help protect and restore the Great Barrier Reef and build its resilience in the face of major threats.

Increasingly their impact reaches further, with solutions developed to address the common challenges facing the world's coral reefs."

Search: Ranger Bot Project

This 'Swiss army knife'-style robo reef protector, the Ranger-Bot Autonomous Underwater Vehicle, will provide reef managers, researchers and community groups extra 'hands and eyes' in the water to:

- control pests like the Crown-Of-Thorns Starfish,

- monitor reef health indicators like coral bleaching and water quality, and

- map expansive underwater areas at scales not previously possible.

Contact:
Great Barrier Reef Foundation
e info@barrierreef.org
t +61 7 3252 7555

office Level 11, 300 Ann Street, Brisbane, QLD Australia

postal GPO Box 1362, Brisbane, QLD 4001 Australia

Let's not forget that any of these actions will also help:
- the winged ones
- the crawlers
- the swimmers
- the hoofed and claw-toed ones
- the microscopic ones

Chapter 12

Ice Sheets Melting

We have covered the disruptions of what lies beneath on levels of climate and social justice in previous chapters. From a climate perspective, let's look at what gives us chills as ice is broken down, sometimes only taking a few days.

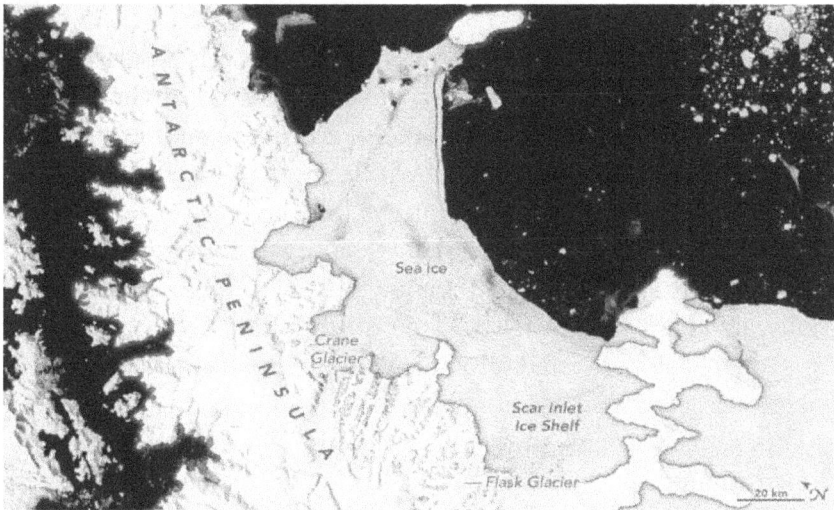

January 16, 2022- Larsen Ice Sheet starting to break along the Antarctic Peninsula, Courtesy NASA Earth Observatory.

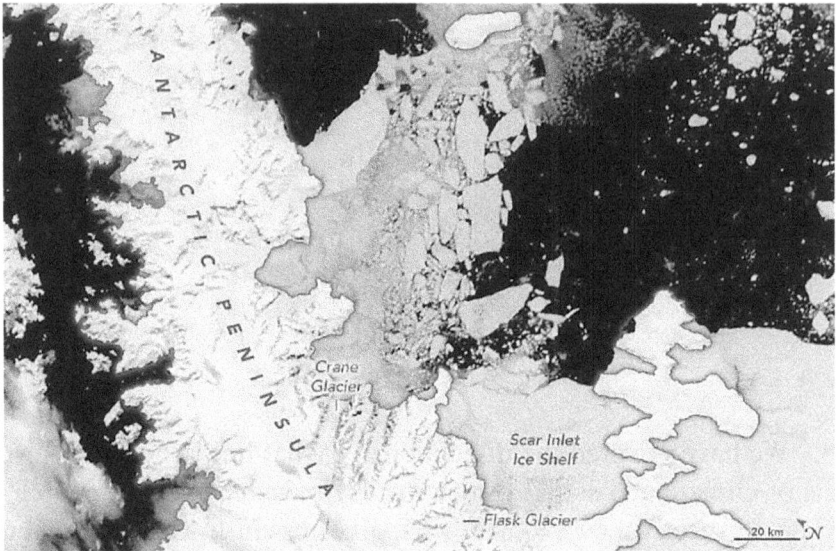

January 26, 2022- Larsen Ice Sheet Breakage along the Antarctic Peninsula, Courtesy NASA Earth Observatory.

According to a recent February 2022 research report by Science Daily, an international team of scientists, led by the University of Cambridge, found that the effect of meltwater descending from the surface of the Greenland ice sheet to the bed -- a kilometer or more below -- is by far the largest heat source beneath the world's second-largest ice sheet, leading to phenomenally high rates of melting at its base. This is caused by huge quantities of meltwater falling from the surface to the base. As the meltwater falls, energy is converted into heat in a process like the hydroelectric power generated by large dams. [1]

As Dr. George Somero, Professor in Marine Science, mentioned in an earlier chapter, when ice melts, the earth is exposed and gets hotter. He says, "take for instance, water melts from glaciers and continental ice sheets particularly where the solid ice is lying on top of land, then melts and floats to the sea causing it to rise. Icebergs make no difference to the rise in sea level because it is not on land. In other words, ice displaces or disperses the same volume of water as it produces when it melts. This can create serious hazards to shipping vessels however where these floating icebergs may not be seen until there is a ship wreck. We are acquainted with this from the famous Titanic."

Do note that ice sheets can cover oceans and land and glaciers cover mountains.

This can also make one think about melting ice and sea-level rise. According to Dow and Downing, "melting of floating sea sheet ice and calving of glaciers (*Ice calving, also known as glacier calving or iceberg calving is the breaking off of chunks of ice at the edge of a glacier says Wikipedia.*) into the ocean, do not affect the sea level. However, thinning and retreat of glaciers on land does add water to the oceans. The ice sheet covering West Antarctica rests on rock that is below sea level. Were it to break up, sea levels would rise substantially?" [2]

Evidence of such calving is depicted by the iceberg breaking from the Larsen C area ice shelf in Antarctica in July 2017. The iceberg is likely to be named A68 and is said to weigh more than 1 trillion tons and cover 5,800 square kilometers (2,240 square miles). The iceberg is one of the biggest ever recorded and is also said to be about the size of Delaware. [3]

As we look at Antarctica, we note that it is covered by gigantic ice sheets up to 2.8 miles thick (4.5km) and covered by an area of 5.4 million square miles (14 million sq. km). It consists of the huge east Antarctic ice sheet, east of the Transantarctic Mountains, which seems to be gaining ice, and the smaller west Antarctic ice sheet which is losing ice. The ice is melting faster on the Antarctic Peninsula, where temperatures are rising more rapidly than anywhere else on Earth, by up to 5.4^0F (3^0C) since 1951. Dow & Downing further report in their Atlas of Climate Change, from 1950 to the year 2000, average temperatures increased by 2.5^0C- four times the global average. 87% of glacier fronts have retreated.

Sea level rise from melting ice sheets in Antarctica and Greenland is a threatening catastrophe for coastal cities within decades unless strong measures are taken to reduce CO_2 emissions from the use of fossil fuels, argues climate scientist James Hansen. This conversation is chiefly about the areas of Antarctica and Arctic because that is where the world's biggest ice sheets exist.

Speaking of Larsen, A and B areas, the collapse of the Larsen A Ice Shelf in 1995 was followed by that of the Larsen B Ice Shelf in 2002 and showed more disintegration. In 2015 a NASA earth topic study lead by Khazendar, concluded that the remaining *Larsen B* ice-shelf will disintegrate by the end of the decade, based on observations of faster flow and rapid thinning of glaciers in the area.

Khazendar indicates that located on the coast of the Antarctic Peninsula, the Larsen B remnant is about 625 square miles (1,600 square kilometers) in area and about 1,640 feet (500 meters) thick at its thickest point. Its three major tributary glaciers are fed by their own tributaries farther inland. Khazendar also noted his estimate of the remnant's remaining life span was based on the likely scenario that a huge, widening rift that has formed near the ice shelf's grounding line will eventually crack all the way across. The free-floating remnant will shatter into hundreds of icebergs that will drift away, and the glaciers will rev up for their unhindered move to the sea. "What is really surprising about Larsen B is how quickly the changes are taking place," Khazendar said. "Change has been relentless." 4

Looking at the Arctic region, the majority of the Arctic region is covered by huge ice sheets. More than 1.9 miles (3km) thickness of ice is at the center of Greenland.

By the year 2100, most scientists think that sea levels will rise by nearly four feet. Taking into account the ice sheets of Antarctica & Greenland we have to look at what actually affects rising sea levels.

For an interactive map to see threats from Antarctic ice loss, please search:

Climate Central Ice Loss Map 5 and select Ice Sheets.

POLAR BEAR CONCERNS:

Who could write about melting ice sheet without mentioning Polar Bears?

There are other animals in the Arctic region; for example, to name a few–moose, wolves, owls and a various birds and sea animals -whales, seals, sharks, walrus, sea otters. This is not to exclude some in the surrounding ocean of Antarctica such as penguins, seals, whales and birds, crustaceans; namely crabs, sea spiders and shrimps as well as species of fish. The most talked about is the saga or story of the polar bear.

If the warming of the Arctic region interferes with their food chain, polar bears will suffer most because it takes a huge number of crustaceans (crabs, lobsters, crayfish, shrimp, krill), fish and seals to support just one bear. Another problem with the warming temperature is vanishing ice. Since a polar bears specialty is hunting on ice, their ability to hunt when ice disappears could mean extinction. They have to swim very long distances between areas of unstable ice. When summer comes, the ice melts even earlier. This means the polar bears start coming ashore. Often, they have not eaten enough reserves to allow them to survive without eating until the sea freezes again.

So, whether the sea levels rise or the ice shelves continue to collapse and break, a warming ocean could be very destructive.

Looking into the mirror where humans are concerned, the melting effects of social justice have taken place along with the reconstructing and altering climate. For example, this includes the indigenous folks of the Arctic regions as Greenland, the Inuit as well as people of color in the United States of America. The consequences are grave and will continue to grow as warm ocean currents accelerate movement. Listening to the wisdom, of Angaangaq (a-rawn-), a renown elder of the Arctic region, "We are reminded of the inevitable destruction that will come about from the melting ice at the top Arctic area down to surrounding countries and cities that will be problematic as the ocean rises with increasing waves from storms. Those countries who will suffer are India, Bangladesh, Cambodia and Vietnam. As most people live in coastal areas, big cities in North America such as New York City, Philadelphia, Houston, Los Angeles will also suffer, whereas cities in Italy such as Venezia could be gone if there is nothing done about it." 6

As we speak or shall I say as we read, conditions involving permafrost also must be considered, for here is where many are being deprived of homes and economy. A good definition other than frozen earth comes simply from Wikipedia: Permafrost is ground that continuously remains below 0 °C (32 °F) for two or more years, located on land or under the ocean. Most common in the Northern Hemisphere, around 15% of the Northern Hemisphere or 11% of the global surface is underlain by permafrost. This includes substantial areas of Alaska, Greenland, Canada and Siberia. It can also be located on mountaintops in the Southern Hemisphere and beneath ice-free areas in the Antarctic. It can be from an inch to several miles deep under the Earth's surface. It frequently occurs in ground ice, but it can also be present in non-porous bedrock. Permafrost is formed from ice holding various types of soil, sand, and rock in combination. 7

According to NOAA's recent 2021 Arctic Report Card: "About five million people live in the Northern Hemisphere permafrost region, which includes glaciers, and within this region, glacier and permafrost hazards are affecting lives, infrastructure, and ecosystem services. Recent degradation of glaciers and permafrost in the Arctic are leading to emerging biogeochemical threats that have the potential to disrupt ecosystem function and endanger human health (Miner et al. 2021). Thawing of ice-rich permafrost can cause ground subsidence with negative implications for infrastructure, ecosystems, and human lives and livelihoods (Suter et al. 2019; Gibson et al. 2021), while even a warming of permafrost can cause a reduction in its bearing capacity, impacting its ability to support structures. 8

Consider the commercial fishing industry, a coastal source of income, when there is no way to hunt and fish. Thus, the indigenous people's livelihoods are affected.

Additionally, as the climate changes, it continues to affect the animals. Just think of the polar bears for instance being spared only small bits of ice to play and fish. Is this social injustice or climate injustice for animals? In any event, the melting ice from a social justice viewpoint is indicative of what gives us chills as systems breaks down. Reflected in the mirror of humanity, yet true to the melting-away metaphor, the actually seeing and witnessing of the issues that had been hidden to blind eyes. This was brought out by the coronavirus affecting African Americans and Latinx populations.

What was broken like the ice chunks at sea was and is the health care system and all other related systems that involved and still involves many African Americans, the vulnerable and the poor.

Going inside those affected, there are inner-dialogues from many stories that can be summed up into one conglomerate story:

"Hold off with this sickness I feel. I don't have the medical insurance to pay for it and my job does not pay enough for me to see a doctor and also pay my rent. That place I call home is pretty run-down but the rent is cheap where us black and brown folk have to live in my very poor-folk neighborhood. I can't afford to lose what little I make by getting fired for not showing up. Besides, me and my kids will starve and be put out on the streets if I take time out to deal with my sick body. I can't even go check on my sugar problem and I want to lose some of this nearly 200-pound body. Then, if I have to go to the hospital, who's going to care for my kids. Just thinking about what some people say, I never gave it a thought that riding the city transit to work and other places was killing me with the virus lurking and nobody knew to wear a mask then. And not even the public knew better as I checked-out groceries where I worked."

In other words, the broken-down systems reflect some issues as follows-

- Death toll of African Americans and people of color from coronavirus

- Employment concerns (not enough wages, no sick leave, no insurance)

- Housing (living in sub-standard housing that is often not safe

- Food Stability (wanting nutritious food from supermarkets or farmer's markets and not having to rely on convenience stores that sells foods with lots of high-fructose, fat, salt and artificial ingredients)

So, with surmounting death toll of Black bodies, of elderly bodies, of Indigenous person bodies and others including the melting earthly ice, the sadness drifts in broken pieces that should melt the heart of any compassionate soul. The next step is taking action to help alleviate these issues.

ACTION

THE NEW STORY OF HEALING & SOLUTIONS

Every solution and resource births a new story for change. If we want to see changes, we must take action.
In this part of the chapter are possible solutions and resources to help inspire actions so we can have opportunities to

A) Live new or revived stories to: awaken to the conditions and consequences that are risky where people and animals live and get involved in projects and careers to help reduce the greenhouse gases, benefitting both people and planet.

B) Find ways to dissolve systems that eliminate care, compassion and proper housing for BIPOC and vulnerable populations and create new structures that benefits all human being

C) Find any action item as a pathway towards a climate career, social justice career, research or volunteer project that is just waiting for you to choose from. That is, unless you have already been triggered into action.

List of possible action items to pursue and get involved with as concerns Melting Ice to help people and animals living in those areas:

1. Polar Bear International [9]

info@pbears.org
Polar Bears International—US
PO Box 3008
Bozeman, MT 59772

"Our mission is to conserve polar bears and the sea ice they depend on. Through media, science, and advocacy, we work to inspire people to care about the Arctic, the threats to its future, and the connection between this remote region and our global climate."

Research:

Population studies track changes as the polar bear's habitat shrinks. This research provides governments and management authorities with critical data for making decisions—helping to conserve the bears. Funding is cureently provided for studies in Western Hudson Bay and Southern Hudson Bay.

Den Studies: This research documents key aspects of denning behavior and adds to our knowledge of den habitat selection. The findings will help managers and policy makers establish the best possible guidelines for assuring the welfare of denning bears. It will also help scientists understand the impact of climate change on the critical reproductive function of denning. Studies are currently being conducted in Alaska and Svalbard.

2. Alaska Conservation Foundation [10]

info@alaskaconservation.org
1227 W. 9th Ave., Suite 300
Anchorage, Alaska 99501

"Although remote, the Arctic is under constant threat from the massive reserves of oil, coal, gas and minerals housed underground. It is also the poster child for climate change. Given the sensitive ecosystems and short summer growing periods, the Arctic is incredibly susceptible to destruction and devastatingly slow to recover."

"The last 50 years, average temperatures in the region have increased 4 degrees Fahrenheit — which is four times the global average temperature increase. The polar ice cap has shrunk nearly 40% since 2005. Permafrost has receded by nearly 10%. Sea ice is declining at a rate of 13.2% per decade."

"Today, the Foundation provides operational support to groups organizing in the region, including Gwich'in Steering Committee and Northern Alaska Environment Center, and provides project funding to keep the Arctic region in the national spotlight. Alaska Conservation Foundation's Alaska Native Fund has also helped empower Alaska Native groups to organize and find financial support."

"Our grantmaking seeks to build the influence of Alaska's conservation movement, focusing upon long-term, enduring solutions that create more robust environmental policy and protections."

3. Climate Justice Resilience Fund:
1201 Connecticut Ave, NW Suite 300
Washington DC 20036 11

"Climate change is altering the Arctic faster than any other region. In our funding areas of Alaska, Canada and Greenland, profound impacts are well under way: melting tundra, shrinking sea ice, lack of snow, and rapid shifts in native mammal, fish, and plant populations. Indigenous peoples face identity, cultural and livelihood disruptions, since their language, traditional foods, and way of life all rely heavily on Arctic land and waters. For example, hunters and fishers now face risky travel across melting ice and tundra to pursue animals with new behaviors and migration patterns. Some coastal communities face forced relocations as the melting land erodes out from under them. And, as the ice recedes, rapid industrial development puts additional pressure on local resources, with multinational corporations and other outsiders taking advantage of new shipping routes and previously inaccessible mineral, oil, and gas deposits"

"CJRF's grantmaking supports Indigenous communities to adapt to this very complex context. We use a framework based on transformative change, which means fundamental change a system. The Arctic faces several climate-driven transformations in ecological systems, such as the likely loss of summer sea ice, and the shift of some tundra areas to forest. But positive social transformations are also possible: consider, for example, the end of Apartheid in South Africa, or the rapid movement out of poverty by many East Asian countries. We have established four regional objectives aimed at supporting Arctic Indigenous peoples to optimize their own societal transformation as the environment around them shifts:"

1. Prepare for the Journey:

Outdated regulatory frameworks and the settled nature of formerly nomadic communities have created new barriers that prevent Arctic peoples from bringing their traditional resilience fully to bear on today's challenges. Our first objective, therefore, is to support communities in having voice and control in the initial steps of change: defining problems, envisioning solutions, and tracking progress. Illustrative outcomes include:

- Effective Indigenous engagement in climate-related decision process

- Development of Indigenous-owned adaptation planning and monitoring tools

2. Charting a Path

The road to transformation often has surprises and unexpected turns. A profound crisis may even offer an opportunity to initiate positive change. CJRF's second objective seeks to support communities in developing flexibility, transparency, interconnectedness, and other strengths for navigating the unpredictable, in the context of two outcome areas:

- *Climate-forced Displacement:* We support rights-based, community led approaches to relocation that safeguard life, livelihoods, community integrity, and self-determination.

- *Social Movement Infrastructure:* We support strong Indigenous climate advocacy coalitions and structures to support exchange and peer learning among adapting communities.

3. Build Strength

Movements are getting people back on the land and renewing traditional language and culture. These social developments could build momentum as part of a resilient response to change, and have a lot to offer the process of transforming the current paradigm. With this in mind, CJRF supports initiatives that:

- *Maintain, update, and augment Indigenous knowledge and ways of knowing*

- *Adapt and sustain traditional livelihoods, resource stewardship, and wild food access*

4. Engage Audiences

A societal transformation is unlikely to take place one isolated village at a time. Communities will need to collaborate and exchange lessons, and take part in decisions that cut across local, regional, national and international levels. This presents important challenges in a region where people are thinly dispersed across a huge landscape, far from national and international centers of power. In this context, strong, creative, and consistent communications become an essential ingredient for change. CJRF aims to empower Indigenous Arctic communities to engage audiences in their work by building communications capacity within the non-profit sector. Example outcomes of work under this objective might include:

- A cohort of young Indigenous leaders with strong communications skills and clear narratives around social justice approaches to adaptation and resilience

- New, more effective climate resilience communications strategies, narratives, or tactics shared among advocacy networks and alliances

- Improved communications systems for knowledge exchange among communities actively grappling with he impacts of climate change.

5. Listen to the Elders Speak

Listen to wisdom spoken about Greenland's melting ice, a video produced by Ice Wisdom or search for: "Angaangaq about the Conditions of the Environment." 12

Suggested Actions to Find Solutions and take Action when Systems are Broken:

- Employers look to fill positions-rank candidates based on performance rather than identifying factors such as their name, race, where they went to college, etc.)

- Get your community involved.

- Find effective strategies that end racism and promote racial healing.

- Provide participants with a cognitive and affective understanding of the cultural, structural, institutional, and political aspects of racism and the development of skills in working effectively.

- Involve everyone in the community to have dialogues (black, white and all persons) that includes all groups – not just people of color – and use a community facilitator to discuss explicit framework of how structural and personal racism creates and keeps sustaining the problem that still maintains inequities and find ways to resolve them.

- Participate in empathy awareness work through Bridge Building to Equity Workshops-
https://www.lavernemcleod.com
info@lavernemcleod.com

Let's not forget that any of these actions will also help:
- the winged ones
- the crawlers
- the swimmers
- the hoofed and claw-toed ones
- the microscopic ones

Hydroelectric Power Plant

Chapter 13

Water As Renewable Energy

We have to consider using more hydroelectric power as water is a valuable and renewable resource. This is mainly because the water cycle is an endless recharging system and is considered a renewable energy. When flowing water is captured and turned into electricity, this power emerges.

Water constantly moves through a vast global cycle as it evaporates from lakes and oceans, forming clouds, precipitating as rain or snow, then it flows back down to the ocean. The energy of this water cycle, which is driven by the sun, can be tapped to produce electricity or for mechanical tasks like grinding grain. Hydropower uses a fuel—water—that is not reduced or used up in the process.

There are several types of hydroelectric facilities; they are all powered by the kinetic energy of flowing water as it moves downstream. Turbines and generators convert the energy into electricity, which is then fed into the electrical grid to be used in homes, businesses, and by industry. [1]

According to the Wisconsin Valley Improvement Company,

- Hydropower is clean. It prevents the burning of 22 billion gallons of oil or 120 million tons of coal each year.

- Hydropower does not produce greenhouse gasses or other air pollution.

- Hydropower leaves behind no waste.

- World-wide, about 20% of all electricity is generated by hydropower.

- Hydropower provides about 10% of the electricity in the United States.

- The United States is the second largest producer of hydropower in the world. Canada is number one. [2]

This form of renewable energy can provide power for isolated places as well as provide power into the electricity grid. The running water flow must be strong enough to build up enough water pressure and to keep the turbine running during dry weather. This may involve digging a small mill pond if there is the space to do so. When dams have to be built to store water, sometimes wildlife cannot move about easily and sometimes silt builds up. A small-scale hydro system does not need artificial reservoirs, thus no release of methane from decaying biomass. Global warming is enhancing the hydrolic cycle: more evaporation, which leads to more precipitation. Whereas, flooding is a threat from this increased precipitation, more water flow for driving electricity generation is at least a minor silver lining in the dark cloud. Thus, the capturing of flowing water being turned into electricity is mighty powerful, hydro powerful.

Resources:

Explore how different hydro power companies operate and

perhaps you will find your career or job opportunity. Search: "The Top 10 Hydropower Companies" 3

Search: hydropower manufacturers. 4

The Elemental Mirror of Earth

Earth/Soil-sprouts

What's on Earth?
What's in the Earth (dirt) where we grow most of our foods?

What's happening with Earth in our overly-heated environment that affects both planet and people particularly struggling people and people-of-color)?

What's in Earth's environment with growing food that deletes equity for struggling people and people-of-color?

Here's what the late West African writer and teacher, Malidoma Patrice Some, once said about Earth: *"Earth is where we belong. She is our home. She gives us sustenance unconditionally and makes it possible for us to feel connected. Earth is where we go to and where we come from. The nourishment and support of the Earth Mother grant us the feeling of belonging that allows us to expand and grow because we feel strong."*

Even if we don't know all the answers to any questions right now, let's open up to some of the issues and solutions presented in the next four chapters of this section. Please commit to peacefully act on one or some of them to help create a better world!

Dryness

Chapter 14

Heatwaves And Drought

We have reflected on watery to icy situations in our environment, both socially and climate-wise. Let's give some time to touch on the story of drought, one of the causes that can eventually lead to fiery circumstances. As we do this, let's note that even though this chapter addresses the condition of drought, it also affects all humans and living matter on Earth.

As we know, drought is defined as an extended time when a region receives a deficiency in its water supply, whether atmospheric, surface or ground water. This drying-out effect is based on many sources. Some scientists call it desertification or becoming a desert where a dry land region becomes increasingly arid, typically losing its bodies of water as well as vegetation and wild-

183

life. 1

An example is the Amazon River's extreme drying-up spell in 2005. Reduced rainfall was the cause, thus exposing areas of dry cracked mud and dead fish. Those who live on the fringes of the desert are used to drought by relying on seasonal rains for themselves, their crops and animals. If the rain doesn't fall, crops & animals die, plus famine is a threat. This happened in 2006 in southern Ethiopia.

Perhaps not all deserts are caused by lack of rainfall. As some can be attributed to poor farming practices when farmers divert water from a drying land into their irrigation systems.

Another factor to dried up land from farm practices has been studied by Dr. Vandana Shiva in South India in 1984. She observed that there was drought in areas that get the same amount of rainfall each year and that the grazing farm animals were dying. She found the cause to be a new modified seed that had been introduced to increase the sorghum crop yield that had synthetic fertilizer built in. This seed with fertilizer reduced plant height, did not produce straw for animals to graze on and then return their waste to the soil as a natural fertilizer. Not to mention certain losses, that being loss of living organisms, microbes and the worms. Thus, nitrous oxide, a greenhouse gas is emitted in the air, that otherwise would have been naturally sequestered or naturally absorbed by organic material. Then problems arise. Nitrates flow into drinking water, in rivers and eventually to the ocean.

More nitrogen in the water creates more algae that die and sink to the bottom. While these plants are de-composing, the oxygen demand goes up. So, all of a sudden where there was oxygen, there isn't any and fish and other animals near the bottom die. There are areas in Gulf regions like the Gulf of Mexico called Dead Zones where oxygen loss has taken place as well. Dead Zones are increasingly common where rivers enter the ocean around the world. They have been reported as covering about 7,000 sq. miles (about the size of the state of New Jersey). 2

Combining drought with rising temperatures that lead to heat-waves, the issue becomes more complex. In a heat wave, it is obviously very hot and is associated with continuous or long periods of sustained high temperatures. What we see in the mirror, as relates to social injustice due to climate conditions, is that people who cannot afford an air conditioner or other cooling systems, nor resources to leave these hot areas, suffer the most. This is a worldwide concern.

By midcentury, if greenhouse gas emissions are not significantly curtailed, according to Environmental Research Communications, the coldest and warmest daily temperatures are expected to increase by at least 5 degrees Fahrenheit (F) in most areas, rising to 10 degrees Fahrenheit by late century. A recent study projects that the annual number of days with a heat index above 100 degrees F will double, and days with a heat index above 105 degrees F will triple, nationwide, when compared to the end of the 20th century." 3

There was a European heat wave in 2003. Portugal, Prague and Paris suffered with nine days in a row with temperatures above 95°F which is 35°C. When hot days and nights continue, it's hard for people to cope with the heat. Also, the added stress on the air conditioning systems can cause them to wear out easily and quickly or even stop them from working. The most vulnerable are the elderly because their bodies cannot lose heat easily. As a result, wherever there's a heatwave, be it in Chicago in 2006 or in Rome, it took a lot of casualties of people over 85 years of age.

Now, what if there are no trees and vegetation to reduce heat in some BIPOC communities, no porous pavement, or affordable cool roofing products? Take for example, a Chicago study found that increasing tree cover by 10 percent could lower total heating and cooling energy use by 5–10 percent annually ($50–$90 per dwelling unit). Case in point, back in 1995, Chicago experienced an extreme heat event that led to the deaths of several hundred people over the course of five days. 4

Climatological studies suggest Chicago will be 5 to 9 degrees warmer by 2100, with some projections of between 9 and 13 degrees, according to data compiled for the Chicago Metropolitan Agency for Planning. Chicago can swing from comfortable to sweltering at the peak of summer. 5

Over a six-day period during the middle of June 2021, a dome of hot air hovered over the western United States, causing temperatures to skyrocket. From June 15-20, all-time maximum temperature records fell at locations in seven different states (CA, AZ, NM, UT, CO, WY, MT). In Phoenix, Arizona, the high temperature was over 115 degrees for a record-setting six consecutive days, topping out at 118 degrees on June 17. 6

Also, in 2021, Europe's heat waves beamed with the hottest on record. Temperatures rose above 40^C or 104^F in parts of Greece and most of the region. Other areas suffered as well across Europe. 7

According to Dow and Downing, heatwaves in Europe, may be linked to global ocean circulation. Across the United States, the EPA (Environmental Protection Agency)'s Climate Change Indicators points out the following:

- Heat waves are occurring more often than they used to in major cities across the United States. Their frequency has increased steadily, from an average of two heat waves per year during the 1960s to six per year during the 2010s.

- In recent years, the average heat wave in major U.S. urban areas has been about four days long. This is about a day longer than the average heat wave in the 1960s.

- The average heat wave season across the 50 cities in this indicator is 47 days longer now than it was in the 1960s. Heat waves that occur earlier in the spring or later in the fall can catch people off-guard and increase exposure to the health risks associated with heat waves. 8

According to the Center for Climate and Energy Solutions, "over the past decade, daily record temperatures have occurred twice as often as record lows across the continental United States, up from a near 1:1 ratio in the 1950s. Heat waves are becoming more common, and intense heatwaves are more frequent in the U.S. West, although in many parts of the country the 1930s still holds the record for number of heat waves caused by the Dust Bowl and other factors.

Also, extreme heat causes more deaths than any other weather-related hazard—more than hurricanes, tornadoes, or flooding, and an average of more than 65,000 Americans visit emergency rooms each summer for acute heat illness" 9

According to EPA (US Environmental Protection Agency), classifying a death as "heat-related" does not mean that high temperatures were the only factor that caused or contributed to the death. Pre-existing medical conditions can significantly increase an individual's susceptibility to heat. Other important factors, such as the overall vulnerability of the population, the extent to which people have adapted and acclimated to higher temperatures, and the local climate and topography, can affect trends in heat-related deaths. Heat response measures, such as early warning and surveillance systems, air conditioning, health care, public education, cooling centers, infrastructure standards, and air quality management, can also make a big difference in reducing death rates. For example, after a 1995 heat wave, the city of Milwaukee developed a plan for responding to extreme heat conditions; during a 1999 heat wave, heat-related deaths were roughly half of what would have been expected. 10

ACTION

THE NEW STORY OF HEALING & SOLUTIONS

Every solution and resource births a new story for change. If we want to see changes, we must take action.

In this part of the chapter are possible solutions and resources to help inspire actions so we can have opportunities to live new or revived stories to:

A. Help cities develop plans to respond to extreme heat conditions so the poor and vulnerable populations do not have to suffer when its extremely hot.

B. Find any action item as a pathway towards a climate career, social justice career, research or volunteer project that is just waiting for you to choose from. That is, unless you have already been triggered into action.

Possible Action Items to Pursue for Dealing with Extreme Heat Conditions in Communities

Consider:

- "Cool roofing products are made of highly reflective and emissive materials (often light colored) that can remain 50–60 degrees cooler than traditional materials during peak summer weather. About 60 percent of urban surfaces are covered by roofs or pavement, traditionally made of dark materials with low solar reflectance (5–15 percent) that absorb about 90 percent of the sun's energy, transferring that heat energy to the ground or buildings below. Cool roof materials have higher solar reflectance (more than 65 percent) and transfer less than 35 percent of the energy to the buildings below them."
"By reducing indoor air temperatures, cool roofs can contribute to lower rates of heat-related illnesses and mortality, especially in homes without air conditioning and in top floors of buildings"

- "Conventional pavements in the United States are made with impervious concrete and asphalt, which can reach peak summertime surface temperatures of 120–150 degrees because of lower solar reflectance (about 5–40 percent). Various types of cool pavement materials have been developed that have higher solar reflectance. Some are also permeable, allowing for more evaporative cooling of pavement surfaces."
"Some cool pavements can also be permeable, allowing air, water, and water vapor into small gaps in the pavement. These pavements address local flooding and urban storm water issues by allowing water to pass through the voids and into the soil or supporting materials below. Some permeable pavements contain grass, which both

absorbs water and is cooler than dark pavement options"

"Trees and vegetation can reduce heat by shading buildings, pavement, and other surfaces to prevent solar radiation from reaching surfaces that absorb heat, then transmit it to buildings and surrounding air. A number of studies have quantified the cooling effect of urban vegetation."

Search for: The cooling effect of urban green vegetation in different climate zones. 11

Preparedness:
Search for EPA Chicago Adapts to Improve Heat Preparedness. Find out what has been done successfully. 12

After the Chicago disaster in 1995, the city put in place together with the Field Museum (a world-renowned museum and a leader in science education and engagement) the development of an outreach program that targeted neighborhoods vulnerable to climate change. This partnership engages the local community to increase awareness of neighborhood vulnerability and identify how residents could reduce the impact of climate exacerbated extreme heat events. This outreach effort has worked to complement traditional disaster response actions that Chicago promoted after the 1995 heat wave. Here is what happened:

Identified current disaster response needs for extreme heat resiliency
- o Chicago adopted disaster responses including: expanding *Notify Chicago,* the city's text and email emergency notification system; identified (e.g., public libraries) or established cooling centers; set up a call "311" program to have officials conduct well-being checks for those who may need additional assistance during events such as the elderly or infirm; and conducted disaster preparedness and response trainings.

- **Assessed citywide vulnerability to future climate**

189

extreme heat conditions

- ○ The city assessed future vulnerability using an "analog city analysis".
  Search for: <u>USGCRP's Climate Change, Heat Waves, and Mortality Projections for Chicago</u> 13

- ○ Chicago identified the most vulnerable residents in their community (e.g., elderly, young) in order to best target heat outreach efforts.

- **Adopted adaptation strategies that targets extreme heat vulnerabilities and support vulnerabl populations**

  - ○ By partnering with the Chicago Field Museum to conduct tailored outreach to at-risk communities and neighborhoods in order to increase understanding of current and expected future changes. Relating future climate norms to a previous extreme heat event (in this case, the 1995 heat wave) helped the climate risk resonate with the residents. Chicago identified urban heat island areas that would be worsened by climate change and used this information to target green infra structure and heat island mitigation efforts.

Other possible ways communities can be prepared for extreme heat events:
Search for Extreme Heat Event Guide 14
Even though information is provided by several agencies to public officials for community interventions for EHE (extreme heat events), it is often not adhered to in BIPOC communities. This information is provided by NOAA (National Oceanic and Atmospheric Administration), CDC (Center for Disease Control), DHC (Department of Homeland Security, FEMA (Federal Emergency Management Agency). Therefore, the following interventions are imperative to be in place prior to EHE and some communities are

left out of the equation:

- Sending clear public messages emphasizing health protection

- Informing the public of anticipated extreme heat conditions (the danger, how long it will last, how hot it feels at specific times of the day)

- Assisting those at greater risk (those living in nursing homes and public housing, providing homeless intervention services)

- Opening cooling centers to offer relief for people without air conditioning and urging the public to use them

- Providing additional resources of information (toll free numbers, website addresses for heat exposure symptoms and response)

- Opening hotlines to report concerns about individual who may be at risk

- Coordinating broadcasts of extreme heat events response information in newspapers, on television and radio

And the do's and don'ts are also common sense for individuals to follow:

Do:
   o Use air conditioners or spend time in air-conditioned locations such as malls and libraries

   o Use portable electric fans to exhaust hot air from rooms or draw in cooler air

   o Take a cool bath or shower

- Minimize direct exposure to the sun

- Stay hydrated – regularly drink water or other nonalcoholic fluids

- Eat light, cool, easy-to-digest foods such as fruit or salads

- Wear loose fitting, light-colored clothes

- Check on older, sick, or frail people who may need help responding to the heat

- Know the symptoms of excessive heat exposure and the appropriate responses.

Don't:
- Direct the flow of portable electric fans toward yourself when room temperature is hotter than 90°f

- Leave children and pets alone in cars for any amount of time

- Drink alcohol to try to stay cool

- Eat heavy, hot, or hard-to-digest foods

- Wear heavy, dark clothing.

MORE POSSIBILITES TO CONTACT AND BE INVOLVED

1. Center for Climate and Energy Solutions:
Contact for Information: 15
    3100 Clarendon Blvd.
    Suite 800

Arlington, VA 22201
703-516-4146

2. Future Earth Contact for Information: <u>16</u>

"Globally, 2% of total working hours is projected to be lost every year, either because it is too hot to work or because workers have to work at a slower pace. Lost productivity from heat stress at work, particularly in developing countries, is expected to be valued at $4.2 trillion dollars per year by 2030, driving more inequality. The agricultural sector, where 940 million people earn their livelihood, is set to be harder hit by hotter temperatures, pushing workers, crops and livestock past their physiological heat and drought tolerances. This will result in lost labor, in smaller harvests for farmers, higher prices for consumers, and negative impacts on livelihoods"

"Increasing energy demand for cooling also comes as an extensive economic cost to residents, businesses, and governments. Often energy grids are unable to supply the required power for air conditioning in cities during heatwaves. This results in not only increased emissions from carbon-based energy sources, but the failure of power grids, loss of power to businesses, hospitals, and critical infrastructure – compounding loss of productivity, increased costs for the energy sector, and reduced access to life-saving cooling and medical care."

"New initiatives to create early warning and response systems and improved communication approaches can help to save lives and build resilience to heatwaves. Thirdly during a heatwave, decision-makers can employ a range of strategies and policies to modify social behavior and reduce exposure to heat by closing schools or offices which lack adequate air conditioning, ensuring availability of water, health care and first aid, and extending access to pools, parks and public cooling centers. Finally, improved urban design and sustainable planning that increases the amount of and access to green space and other cool environments (pools, air-conditioned spaces) and encourages white roofs, will play an important role in avoiding heat-related illness, in reducing surface temperatures, and in providing a wealth of other nature-related benefits"

3. Heat Shield: 17
Prof. Lars Nybo
Institution: University of Copenhagen
Country: Denmark
e-mail: nybo[at]nexs.ku.dk

"HEAT-SHIELD is a research program funded by the European Union which aims to address the negative effects of climate change, i.e., increasing workplace temperature, on its working population. The effects of heat exposure include productivity loss in many jobs, and HEAT-SHIELD will study this issue and its prevention in different sectors. In addition, heat strain as a result of the increased workplace temperature will pose health issues to the workers of the EU.

To address these issues, HEAT-SHIELD focuses on providing adaptation strategies for the five major industries of the EU and its workers: manufacturing, construction, transportation, tourism, and agriculture. Together, these industries represent 40% of the EU's GDP and 50% of its workforce.

4. **NOAA states**, "Although droughts are costly natural disasters, they are often overlooked when communities plan for hazards. Unlike natural disasters such as tornadoes or earthquakes that inflict dramatic damage in a short period of time, droughts are slow-moving and do not usually cause direct property damage."
There are however steps to consider and take:

- One of the first steps is to define drought for the particular region and identify indicators that describe drought conditions.

- A community's vulnerability to drought is assessed by analyzing historical droughts impacts and regional water use.

- Drought plans can identify actions to be taken when drought occurs, as well as long-term changes to reduce the likelihood of a future drought.

- Proper drought planning includes drought monitoring.

- Early warning systems help communities respond to drought at its onset.

5. A great graphic resource for Drought Preparedness can be found by searching for: Wikihow Prepare for a Drought 18

Let's not forget that any of these actions will also help:
- the winged ones
- the crawlers
- the swimmers
- the hoofed and claw-toed ones
- the microscopic ones

If you live in a drought prone area, what are some things you can do?

Vegetables

Chapter 15

Part I
Food And Farming

Let's think of food and farming as one entity. Because we have to eat, we rely on some type of farming involving growing crops or eating animals to get us there. Of course, those who know me understand that I support the green side of food. I also know that not all of the world eats in this manner.

In this chapter, we are looking into the mirror of social justice and climate justice of growing food. There will be overlapping content as the two are explored. From a social justice mirror we shall cover some of the struggles of farmers of color and the health care disparities involved in feeding their families. On the flip side of the mirror various methods of farming shall be explored in effort to find methods that can effectively and fairly feed all of the planet's people. Suggested solutions and resources that can facilitate both social and climate concerns about farming for food shall conclude this chapter.

Looking through the social justice mirror, one might question; Why have there been struggles for farmers-of-color to grow food? Let's ponder this question as we read the quote:

"The first essential component of social justice is adequate food for all mankind. Food is the moral right of all who are born into this world" Norman Borlaug

It would be beneficial for the planet if complete fairness was granted to all food growers- Black farmers, Indigenous Nations, Asians, LGBTQ, gender and non-gender confirming folks. To start, let's identify some struggles that have hindered the process of fairness towards food growers in the United States. One might bet that this is common in other countries, showing up in one form or another.

The economic inequity part of the US society has to be considered to understand what is currently happening. By understanding that and a rigged past, we can better understand the present. Solana Rice, Co-founder and Co-Executive Director of Liberation in a Generation, lays it on the line in her talk about economic inequity and shameful wealth gap that is problematic. She says, "for people of color in the United States, the root of financial insecurity stems from institutional racism and white supremacy that has existed since the founding of the country." She states that "home ownership [and other ownerships of land] where people of color were systematically denied, started in the 1930's with the inability to purchase homes because of the racist scheme called redlining. The color "red" was used to show black and brown neighborhoods on city and county maps. This resulted in refusal to sell homes in white neighborhoods to BIPOC. Thus, as of 2019, 60% of redlined communities remain low income neighborhoods." A recent example of these practices was the sub-prime mortgage disaster that was pushed on unsuspecting BIPOC people in 2008. 1 All of this reminds of the Black Codes and Jim Crow Laws, as a continuing of suppression of BIPOC.

From a historical perspective, the reflective mirror left little room for hope for BIPOC to grow food. In spite of this, a type of resilience remained throughout the history of farm struggles to own farms. Here are some of the ups and downs of those struggles:

- **Down**- Stealing of Indigenous Nations land that started with the Papal Bull of 1493. A papal bull is a type of public decree, letters patent, or charter issued by a pope of the Catholic Church. It is named after the leaden seal (bulla) that was traditionally appended to the end in order to authenticate it. [3]
The Papal Bull of 1493 written in Latin was issued by Pope Alexander VI of Spain: giving exclusive rights to the New World that Columbus said he discovered in 1492 with certain powers: The power to ENSLAVE or KILL NATIVE INHABITANTS & claim the land. This weighed heavy on the fact that there is very little that Indigenous Nations can still have that was all once theirs, due to treaties and other laws. In the late 1700's and most of 1800's Indigenous land was taken. [4]

- **Down**- In 1783 the US seized 1.5 billion acres of land over the next century after the Revolutionary War. [5]

- **Down**-In 1830 The Indian Removal Act allowed the government to seize lands of Native people in the East and the South in exchange for a harsh colonization zone west of the Mississippi River. Then next came the Trail of Tears that forced migration to the land set up for them. [6]

- **Down**-In 1848 the lower Rio Grande Valley land was taken as Anglos began squatting on the land of Mexican subsistence ranchers 7

• **Down**-In 1850 in California, the state passed a law to remove Native Americans from their lands, separate children from their families, strip people of their cultures and languages, and create a system of indentured servitude 8

• **Up**-In 1862 the Homestead Act passed by Congress allowed citizens to claim 160 acres for a small fee-taking about 246 million acres in the West. This did very little to help the farmers of color and Indigenous people. 9

• **Down**- In 1619, over 400 years ago the first enslaved Africans were brought to what is now the state of Virginia. They were locked in heavy iron chains and crammed onto ships for a dangerous journey- many dying in route from heat, starvation, thirst, and violence. They were brutally branded and marked and forcibly worked the land and homes of white people who owned them like cattle. 10   He 400t

• **Up**- in 1865-Only for a short time, the enslaved Africans were being used also as soldiers in the Civil War (1861-1865). During that time President Abraham Lincoln issued the Emancipation Proclamation in 1863 that all persons held as slaves within the rebellious states shall be free." And January 16,1865, Major General Sherman issued special field orders providing thousands of Black Americans 40-acre plots of tillable land known as **"40-ACRES AND A MULE"** to help recently freed or soon-to-be-freed from enslavement. 11

• **Down-** April 14, 1865, President Abraham Lincoln was assassinated and the 40-acres were taken back or terminated by his successor, Andrew Johnson. Thus, the freed Blacks did not have farmland. This land was to come from white plantation owners so the freed Blacks would have a chance at making a living. 12

• **Down-** In 1871 Congress passed the 2nd Indian Appropriations Act that declared that tribes are not Independent Nations, which allowed more land takeovers. Preceding this, was the first Indian Appropriations Act by Congress that created reservation systems to manage the Indigenous people where they could be subdued or clumped and dispirited. 13

Later, the reservations were divided into individual plots by the Dawes Act in 1887 on mostly unworkable land.

• **Up-** During the Black Freedom Movement of the 1800's some momentum was gained when agricultural scientist, George Washington Carver, pushed crop diversification, composting and other methods to help farmers make enough profit to purchase their land and feed their families. 14

• **Down-** 1882, the Chinese Exclusion Act took effect meaning that economic resentment towards about 15,000 Chinese immigrants took place for those having worked on farms and railroads in the West. This meant that immigration from China was banned and the Chinese workers had to live in urban enclaves. Then the Japanese farmers began to farm. 15

• **Down-** In 1901- President Theodore Roosevelt creates 150 national forests. This meant that Indigenous and Latino communities did not have access to traditional farming and hunting grounds. It was like a starve-them-out and die action. [16]

- **Down**-In tracing immigration history, in 1913, California passes the Alien Land Law which bans the purchase & long-term leasing of land by those "ineligible for citizenship." This Law aimed at farmland operated by Japanese Americans. [17]

- **Up**- By 1920 Black farmers owned a lot of land (15 million acres) by 14% of the US farmers at this time. [18]

- **Down**- In the 1930's, Black farmers were excluded from the New Deal subsidies or endowments by the USDA (United States Department of Agriculture) so white owned farms had the wealth as they were paid to reduce crop production in order to raise crop prices after the Dust Bowl and the Great Depression (1929-1939). [19]

- **Down**-1942 the Japanese attack on Pearl Harbor resulted in Franklin D. Roosevelt signing an Executive Order to force 110,000 people of Japanese descent from their homes and into concentration camps. Thus, many lost their farms and businesses permanently. [20]

- **Down**- More blatant acts of discrimination of Black farmers came from USDA that denied federal aid to Black farmers and forced many to sell or abandon their land. Between 1940 and 1974, the number of African American farmers fell from 681,790 to just 45,594--a drop of 93 percent. [21]

• **Up-** In 1965, the Commission on Civil Rights finds the USDA has discriminated against Black Farmers. As discrimination continued a district court in 1999 in the case Pigford v. Glickman, the USDA were ordered a settlement. Of the 23,000 Black Farmers who filed claims, only 15,645 received payments, mostly $50,000. which was not nearly enough to compensate for the land. [22]

• **Up-** Meanwhile in 2010, some Native tribes asked people to donate a yearly "land tax" as reparations. A Utah woman transferred $250,000 to the Ute Tribe and a farmer in Nebraska signs a deed to return a 1.6-acre plot of native corn to Ponca Tribe. [23]

• **Down-** by 2017, **only 1.4% of all US farmers are Black.** They collectively receive $65 million in annual farm subsidies, while white farmers receive $11.3 billion. [24]

• **Up or Down?** (not yet enacted with the writing) In 2020 the proposed **Justice for Black Farmers Act** is looked at by aiming at correcting "historic discrimination" in federal subsidies and lending that resulted in the loss of millions of acres in farmland and "robbed Black farmers and their families of hundreds of billions of dollars of inter-generational wealth." The bill would devote $8 billion annually to buying farmland and granting it to Black farmers. It is also controversial as Native Tribes want the same for their lost. [25]

In other words, the Justice for Black Farmers Act would enact reforms within the USDA to finally end discrimination within the agency, would protect the Black farmers from losing their land and would provide land grants to create a new generation of Black farmers and being to restore the land base that has been lost by

Black farmers due to discrimination. Specifically, the Justice for Black Farmer Act is a hope to end discrimination within the United States Department of Agriculture. 26, 27

To date, this proposed Act has not been passed.

Inclusive in this struggle are those bearing consequences as agricultural farm workers. Additionally, climate warming consequence as a field worker is of grave concern for agricultural farm workers who are also essential workers.

One such struggle and consequence that presented itself was during "shelter-in-place" in March of 2020 that left high death rates resulting from COVID-19 for essential workers. This health factor noted by University of San Francisco (USF) doctors was virtually every patient in San Francisco's public hospital who was sick with coronavirus was Latino. The reality of health care aides, delivery persons, janitors, cooks, and other essential workers were key to helping the city during lockdown and they were the exposed ones.

According to Stanford University Medicine News Center, a research study team focused on the grim death toll during "shelter-in-place" for coronavirus. They found that Latinos and Blacks populations in California suffered the most.

"Latinos living in California, who are 8.1 times more likely to live in households facing these higher exposure risks than White Californians (23.6% versus 2.9%), had a COVID-19 case rate more than three times that of whites (3,784 versus 1,112 per 100,000 people). Further, California's Latino population was tested for COVID-19 at a lower rate than white population (35,635 versus 48,930 per 100,000 people). The Latino population had strikingly worse COVID-19 mortality outcomes as well: The estimated death rate for Latinos (59.2 per 100,000 people) was 1.5 times higher than white residents (38.3 per 100,000 people)."

Further, "Black populations in California, who also face structural risk factors that put them at high risk of COVID-19 infection and mortality, had the highest death rates (65 per 100,000 people) among the racial/ethnic groups included in the study. During the timeframe of the study, there were 8,635 deaths among Latinos, compared with 5,330 among whites, 2,053 among Asians and 1,295 among African Americans." [28]

Other causes of death by those who need to eat and make a living are agricultural field workers who have to harvest in extreme heat and/or breath toxic chemicals. In current day California as in many states, not all farm hands that plant, tend, harvest and package fruits and vegetables are applauded for their service. They might be considered sacrificial lambs in order that the rest of America can eat. Some farming practices such as using toxic pesticides or toxic fertilizers as well as machinery may be dangerous. This is more common with Latino field workers that are often migrant, undocumented or local residents.

What can be done to improve this system, wherever it exists? Are we thinking humanity and equity in these situations? What about the average Latino farmworker in fields that breathe poisons or pick fruit and vegetables in areas where the temperature is very high 108^{0} and rising with climate warming? In some cases, illness and even death arise. [29]

The same rings true with migrant farmers and harvesters in the Salinas Valley of Monterey County in California, living intergenerationally in households, having had a high statistical COVID-19 rate than other parts of Monterey County. According to the County dashboard tracker, it was reported that Salinas Valley farmworkers during the summertime of 2020, were three times more likely to be infected than other workers at that time. As essential workers, farmworkers planted, harvested and packed produce right beside co-workers, often relying on employers for crowded transportation and accommodation in camp-style housing. In many cases, protective gear, including masks, were in short supply. Testing didn't begin until many migratory workers had moved on to harvests elsewhere. More positive actions will appear in the solutions section of this chapter about agricultural farmworkers.

To continually look into the mirror of food and farming, we can view from the climate justice realm that struggles exist for the soil itself. Since accessing information about destroying the effort for people of color to have farmland, plus the disparities involved with field farmworkers and other essential workers, it is added to the list. There is so much to be learned about farming and this is only a small part that I am including in this book.

So, what happens when soil is destroyed by man, erosion, heat or freezing temperatures?

"We must not only be concerned with what is happening to the soil; we must wonder to what extent insecticides are absorbed from contaminated soils and introduced into plant tissues." Rachel Carson

"A nation that destroys its soils destroys itself. Forests are the lungs of our land, purifying the air and giving fresh strength to our people." Franklin D. Roosevelt

I do not want to think that humans would intentionally destroy the soil upon which they want to raise a crop to produce food. However, without knowing so, many farmlands are tilled and over tilled. According to Crop Management of Iowa State University Extension and Outreach, "tillage has all along been contributing negatively to soil quality. Since tillage fractures the soil, it disrupts soil structure, accelerating surface runoff and soil erosion. Tillage also reduces crop residue, which help cushion the force of pounding raindrops." [30]

Considering a soil focus conducted at the University of Toronto as early as 2008, the question was raised if global warming changes organic matter in soil? Scientists at the University of Toronto Scarborough, researched whether global warming actually changes the molecular structure of organic matter in the soil. Science professors Myrna and Andre Simpson along with Dudley Williams used NMR (nuclear magnetic resonance) facility to gain a detailed view of the soil's molecular structure and reactivity. They also used field work behind the UTSC campus to test the soil with electrodes through summer and winter seasons over a 14-month period to analyze soil samples.

They concluded that "organic matter retains water in the soil and prevents erosion. Also, decomposition of this organic matter provides plants and microbes with the energy source and the water they need to grow. Carbon is thus released into the atmosphere as a by-product of this process." The concern is that warming temperatures may speed up this process. This could increase the amount of CO_2 that is transferred to the atmosphere. 31

Still another question that scientists are concerned with is carbon locked up in the permafrost *(a thick subsurface layer of soil that remains frozen throughout the year, occurring chiefly in polar regions)* in the Arctic. In other words, what would happen when microbes become more active with warmer temperatures?

Well, 12 years later, there is evidence of what the Toronto scientists were concerned about. According to a new study published in February 2020 in the journal, Nature Geoscience, "the ice that holds the soil together is melting, causing hillsides to collapse and massive sinkholes to open up. That dramatic disruption to the landscape is only part of the story — a bigger concern is the growing amount of carbon this process releases into the atmosphere." 32

The rapid melting of previously solid permafrost promotes microbial activity, which releases greenhouse gases like CO_2 and methane into the atmosphere, causing carbon levels to rise. Higher carbon levels, in turn, contribute to global warming. According to a NOAA arctic report, the study warns that the level of emissions from the Arctic could be much higher than previously estimated. 33

Merritt Turetsky, lead author of the study, said the abrupt thawing is "fast and dramatic" and "affects landscapes in unprecedented ways." Other scientists involved in this study are listed in the bibliography.

The scientists summarized the current permafrost thaw saying that: "Active hillslope erosional features will occupy 3% of abrupt thaw terrain by 2300 but emit one-third of abrupt thaw carbon losses. Thaw lakes and wetlands are methane hot spots but their carbon release is partially offset by slowly re-growing vegetation. After considering abrupt thaw stabilization, lake drainage and soil carbon uptake by vegetation regrowth, we conclude that models considering only gradual permafrost thaw are substantially underestimating carbon emissions from thawing permafrost." 34

As we have read what happens with various soils, solid or frozen, we can decide what type of farming we might consider for growing food, should that be our interest. In other words, it may help us determine what track we are on or what we might move away from for growing food. Let's start with the staple food, rice, that most of use eat almost daily.

Rice Farming is a type of farming that seems simple yet its greenhouse gas can create its complexity. According to University of California scientist, Chris Van Kessel, working with the United States Department of Agriculture (USDA), "a significant source of methane emissions and methane (CH_4) from rice farming is a more powerful greenhouse gas than CO_2." He further adds that "methane in rice paddies is produced by microscopic organisms that respire CO_2, like humans respire (breathe) oxygen. More CO_2 in the atmosphere makes rice plants grow faster, and the extra plant growth provides the soil with extra energy." So, methane emissions from rice production will strongly increase as worldwide demand for rice increases.

The United States Department of Agriculture (USDA) indicates that rice, the primary staple for more than half the world's population, is produced worldwide, with about 90 percent grown in Asia. The United States is a major exporter, with the global market accounting for nearly half the annual sales volume of U.S.-produced rice. Four U.S. regions produce virtually all of the country's rice crop—three in the South and one in California—with the South growing mostly long-grain rice and California producing almost exclusively medium- and short-grain rice. [35]

It will be interesting that the USDA has a rice growing research project which is an Agricultural Research Service (ARS) study on water quality with rice growing. This project started September 30, 2017 and is expected to proceed unless further revisions are needed. It will examine Enhanced Rice Pond and Pesticide Simulation features using a computer model tool called the AnnAGNPS (Agricultural Non-Point Source) to simulate or imitate models. This is being done in the Sacramento River basin in California. Other modules such and conditions for rice growing are being studied for Louisiana which includes runoff, sediment, and nutrient components. [36]

The International Rice Research Institute (IRRI) estimated that 10% of global agricultural greenhouse gas emission is from rice production, the second biggest emitter of methane after the livestock sub-sector.

IRRI reports that in the Philippines and Vietnam, rice accounts for 13 and 17%, respectively, of total national GHG emissions. About 60-80% of rice straw is burned in the open field, which contributes to greenhouse emission and air pollution associated with respiratory health problems.

Interesting enough, IRRI states that "rice production is both a victim and a contributor to climate change." This is how they state it:

"Drought, flood, saltwater, and extreme temperatures devastate crops and risk the livelihoods of 144 million smallholder rice farmers each growing season.

At the same time, traditional cultivation methods, such as flooding paddy fields and burning rice straw in open fields, contribute approximately 10% of global man-made methane, a potent greenhouse gas."

The IRRI is not just reporting facts, they are working with governments and other institutions across South Asia, Southeast Asia, and Africa to help find solutions that are responsive to the climate. 37

Despite being a powerful greenhouse gas in its own right that traps even more heat in the atmosphere than methane over long time periods, most rice producing countries do not report their nitrous oxide (N_2O) emissions.

Cattle Farming, is another type of farming that connects with methane due to the flatulent nature of cattle.

According to United Nations Food & Agricultural Organization (UNFAO), cattle farming takes a toll on the atmosphere. This farming method is responsible for 18% of greenhouse gas emissions measured in CO_2 equivalent. This is higher than the transportation industry.

"A cow does on average release between 70 and 120 kg of methane per year. But the negative effect on the climate of methane is 23 times higher than the effect of CO_2. Therefore, the release of about 100 kg methane per year for each cow is equivalent to about 2,300 kg CO_2 per year". 38

Livestock also affect the replenishment of freshwater sources by compacting soil, reducing infiltration, degrading the banks of watercourses, drying up floodplains, and lowering water tables. Livestock's contribution to deforestation also increases runoff and reduces dry season flows.

According to United Nations Food & Agricultural Organization in the United States, livestock are responsible for an estimated 55% of erosion and sediment, 37% of pesticide use (for feed crops) 50% of antibiotic use, and a third of the loads of nitrogen and phosphorus unto freshwater resources.

High numbers of cattle lead to vast quantities of highly concentrated waste. For instance, a typical beef or dairy cow can excrete about 120 pounds of manure per day as much as 20-40 people). 39

A single Concentrated Animal Feed Operation (CAFO), therefore, can produce as much waste as a medium-sized city. The manure is collected in large pools called "manure lagoons," or is applied to fields as fertilizer. Both forms of manure management are known to pollute groundwater through aquifer infiltration and surface water through runoff from over-application of fertilizer and lagoon construction failures. 40

While on the subject of cows causing more environmental issues, let's take a look at the love for hamburgers. According to a Los Angeles Times article, this is something to think about in terms of becoming active to help resolve this matter:

A 1/3-pound **burger** requires **660 gallons** of water. Most of this water is for producing beef.

1 pound of **beef** requires **1,799 gallons** of water, which includes irrigation of the grains and grasses in feed, plus water for drinking and processing. 41

Factory Farming- Raising cattle may seem to be the worst thing in the world considering the damage that can be done to water, land and environment. There is another type of farming that quickly produces quantities of animal food.

Factory farming is very controversial because it deprives livestock, poultry and other animals the space to move about and grow in this mass production format. It helps sellers provide lower costs of food for consumers. This type of farming depends on antibiotics, vitamins, fertilizers and pesticides. This is done to produce more animals and crops rather than the natural conditions normally needed to survive and thrive. In confined spaces, antibiotics help to keep diseases from getting out of control. What does this do to humans who eat excessively antibiotic-dosed meats? Also, factory farms are often located in low-wealth black and brown communities where they pollute the air and water, threaten wildlife habitats, and are thought to be a contributor to greenhouse gas emissions. To add to this, the controversy lies within the confines of cruelty, in cramped tight spaces that are seemingly inhumane for the treatment of animals. 42

With all of this having been said, it is time to dig deeper for solutions.

ACTION

THE NEW STORY OF HEALING & SOLUTIONS

Every solution and resource births a new story for change. If we want to see changes, we must take action.

In this part of the chapter are possible solutions and resources to help inspire actions so we can have opportunities to live new or revived stories to:

A) To partake in positive action to ensure that people of color receive equitable legislative treatment and reparations for farming

B) To learn what foods are healthy to eat, how to prepare them and find ways to balance the social injustice around global food systems.

C) To awaken to the consequences that how we treat farming soil can reduce the greenhouse gases in the atmosphere, benefitting both people and planet.

D) Find any action item as a pathway towards a climate career, social justice career, research or volunteer project that is just waiting for you to choose from. That is, unless you have already been triggered into action.

Although we have examined ways that people of color are struggling to be a part of the farming world to grow food, climate change struggles for the soil itself and controversial farming that damages the environment, livestock and poultry, we still have more to explore. Plus, what we put into our bodies profoundly prepares us with strength and knowledge to do what is needed for issues surrounding food injustices. We have to look at **solutions for food consumption** in relation to the climate crisis from a healthy perspective by linking and harmonizing climate and social justice. To lead us into this solution thought process are two experts from Monterey County in California; Ben Eichorn, "Grow Your Lunch" and Kari Bernardi, "Super Natural Chef."

• Linking the Climate Crisis and Human Patterns of Consumption and Waste

Ben Eichorn, founder of *Grow Your Lunch.*
https://www.growyourlunch.com/ conceptualized on climate and human linkage in relation to food and farming in his abstract:

We Create Successful Organic Farms and Gardens Within Businesses, Schools and Institutions
"According to the United States Department of Agriculture, 30-40 percent of the food supply in the U.S. ends up being wasted." This is an upsetting figure for many reasons. Not only is food not getting to the hungry when an estimated 10 percent of U.S. families experience food insecurity, but all of the energy that went into growing, tending, harvesting, packaging, refrigerating, transporting that food is being wasted, and carbon dioxide and other greenhouse gasses are being emitted for no reason. We are effec-

tively wasting food, while people go hungry, and exacerbating the climate crisis in the process.

It's easy to preach the glories of school and community gardens, more localized food systems and farmer's markets, however, affording locally grown food remains impossible for many Americans, especially for low-income, Black, Indigenous and people of color.

The issue in my perspective is not solely that people don't have access to food that is good for them and good for the environment, it's that we are making an assumption that access is the singular obstacle to achieving food equity in the United States.

The questions we are not considering are diverse and wide ranging. For example, if someone has access to healthy food, do they have the knowledge of how to cook with whole, fresh ingredients? Do they have the appropriate infrastructure at home to cook a meal from scratch (cutting board, chef knife, skillet, stovetop, oven, etc.)? Are the foods available to them culturally appropriate or familiar to them? Do they even have the time to cook while juggling multiple jobs and family obligations? Clearly the issue is not simply one of access. Until every graduating high schooler knows how to cook a meal from scratch, I do not believe we will have food equity in this country.

This is the role that school gardens and culinary arts programs can play in public education. As Chef Alice Waters, founder of the Edible Schoolyard Project in Berkeley, CA, has said, we need a "delicious revolution." It is my hope that we will achieve this vision within the next generation - before traditional food ways, which are inherently better for us and the Earth, are eroded to a point of no return." 43

• Food Choices Harmonizing with Climate Justice and Social Justice

This is what Kari Bernardi, "Super Natural Chef, contributes to the solution: http://www.supernaturalchef.com/

"As a plant-based chef and culinary instructor, I have been sharing and preparing delicious vegan foods with people for over

213

30 years. I have a background of studying environmental and social injustices around global food systems at UC Berkeley where I graduated. Learning about systems where many people were suffering from gluttony and diseases of over-consumption was an imbalance to the system. Also, others were starving and suffering from malnutrition and loss of their offspring due to lack of food and clean water.

Bringing more people to the table to sit and be fed while nurturing and caring for planet Earth, my beliefs are:

a. Eating a diet that consists mostly of plants is an act of environmental and social justice.

b. Plants are here to heal us and provide all the nutrients we need to thrive.

c. All of Earth's inhabitants deserve to be fed and cared for.

d. It is everyone's birthright to have access to fresh food and clean drinking water.

It has become common knowledge that the solution to many of our social and environmental ills can be found on what we put on our plates. Let's consider:

- Limiting our consumption of all animal products, including dairy and eggs, will help us slow down the destruction and possibly turn the tide on global warming.

- Vegan and vegetarian diets can help to secure the world's food supply by slowing climate change from greenhouse gases caused by large scale meat production.
  The 2019 UN Climate Change Report: Food and Land, reports that "an estimated 23% of greenhouse gas emissions come from agriculture, livestock & the land and forests needed to raise them.

- Plants also use less water to produce food and do not pollute water sources like commercial meat facilities do. It takes over a thousand gallons of water to produce 1 pound of meat for consumption and less than 100 gallons to produce 1 pound of veggies. By switching to a plant-based diet we can limit relying on deforestation for cattle grazing and raising and save the Amazon and other at-risk lands needed for carbon sequestering.

Following a plant-based diet daily or a few days a week helps alleviate a myriad of problems. We can choose soil fertility over soil erosion, clean water over polluted waterways, more people fed or more people hungry and plant and animal species survival or species extinction.

I know what I am choosing. Will you join me?" 44

• Possible Solutions for People Equity with Food & Farming

- Immediately **grant reparations to African American people** and for programs to help level the playing field for equity.
  National African-American Reparations Commission, info@reparationscomm.org (718) 429-1415
  Learn about reparation plans and find out which category of reparations you might be called to assist. 45

- Also, **every state governor could apologize to Black leaders** about forcing their ancestors into enslavement. Use the same website above.

- Consider **reparations for Native Americans** to also include water rights. One form of this is with Sonoma County, known as California's wine country, officials agreed in 2015 to transfer nearly 700 acres of the Kashia Band of Pomo Indians' ancestral lands back to the tribe. Do this in various open land spaces. 46

- Read about the protocol applied to Salinas Valley Farmworkers to **close the health disparity for farm workers safety** with coronavirus and its variants.
  by Laura Reiley and Melina Mara. 47

• Possible Solutions for Soil and Animal Species Health

- **Regenerative Agriculture-** A great solution to enrich the soil for growing food.
  According to Treehugger, a source for science and agriculture article, this method of farming is the key to healthy soil. Treehugger defines Regenerative agriculture as a sustainable method of farming that can replenish nutrients in the soil while combating climate change. Regenerative agriculture is a modern name for the way farming was practiced for centuries, before the onset of industrial agriculture in the early 20th century. Returning to those traditional practices is gaining momentum as a way of reversing the damage done to the climate and soil that we all depend on for our food and survival. 48
  A leader in the field of Regenerative Agriculture, Ronnie Cummins is the founder and director of the Organic Consumers Association, a non-profit, US- based network of more than two million consumers dedicated to safeguarding organic standards and promoting a healthy, just, and regenerative system of food, farming. According to Ronnie Cummins, the bottom line is that humans have put too much CO_2 and other greenhouse gases (especially methane and nitrous oxide) into the atmosphere (from burning fossil fuels and destructive land use), trapping the sun's heat from radiating back into space and heating up the planet. And unfortunately, because of the destructive food, farming, and forestry practices that have degraded a major portion of the Earth's landscape, we're not drawing down enough of these CO_2 emissions through

plant photosynthesis to cool things off.

Regenerative agriculture and animal husbandry are the next and higher stage of organic food and farming, not only free from toxic pesticides, GMOs, chemical fertilizers, and factory farm production, and therefore good for human health; but also regenerative in terms of the health of the soil, the environment, the animals, the climate, and rural livelihoods as well. Or as my fellow steering committee member for Regeneration International, Vandana Shiva puts it: "Regenerative agriculture provides answers to the soil crisis, the food crisis, the climate crisis, and the crisis of democracy."[49]

- Read article from Regenerative International "Regenerative Food and Farming: The Road Forward"

03/02/2021/by Ronnie Cummins

To further understand regenerative farming this is how it is done:

A) No tilling or minimal tilling. It protects the microorganisms and bacteria that live in the soil.

B) Plant Cover Crops. According to Trey Hill, a regenerative ag corn farmer in Rock Hall, Maryland, says you never want to see the ground after harvest. Plant Cover Crops (rye, clover, turnips, & other species). They help to keep the soil planted at all times- not just when you are growing food crops. As the winter cover crops grow, they will feed microbes and improve the soil's health, which Hill believes will eventually translate into higher yields of the crops that provide this income: corn, soy bean and wheat.

But just as importantly, cover crops will pull down carbon dioxide from the atmosphere and store it in the ground.

C) Use lots of Compost. Compost is an important part of the cycle of life in the garden. 50

- **SilvoPasture** ("sily" is a Latin prefix meaning "woods" or "tree.") is highlighted in a case study done by Compassion in World Farming. You can view on YouTube silvopasture in action. One can understanding better what this method of cattle farming is all about. It can be summed up as the practice of integrating trees, forage and the grazing on the same unit of land. Multiple components are part of this system. There are trees, forage and livestock crops that helps sustain some of the nutrients. It could be bringing in livestock or other animals and fouls such as duck to the woods or vice versa. Thinning out enough trees to get sunlight to the wood floor so vegetation can grow for the animals to feed on. Timing as to the length of days or months livestock should remain with the trees is crucial to management to avoid eventual damage to the trees, while still providing shelter and shade for the animals during certain seasons.

Here are some beneficial points from Compassion in World Farming as a result of that case study:

"When compared with conventional extensive systems without trees or bushes, the benefits of this system can include:

- More resilience to climatic changes due to the different species having different responses to weather stress. For instance, grasses may be less tolerant of drought than shrubs, so in dry periods, the livestock can eat the shrub leaves rather than the grass.

- Higher diversity of forage, which can improve

218

nutrition.

- Improved animal welfare due to reduced temperature stress, reduced parasite load, reduced stress and increased nutrition.

- Increased productivity due to reduced heat stress and increased nutrition from better distribution and availability of forage throughout the year.

- Increased profits due to reduced input costs and increased productivity.

- Reduced GHG emissions due to reduced fertilizer use. Reduced methane emissions from improved Silvopasture is part of a trend globally to sustainably coax more food from each acre — without chemicals and fertilizers — while reducing greenhouse gas emissions, increasing biodiversity, and enhancing the land's ability to withstand the effects of climate change.

- Better soil water retention and higher water infiltration to deeper soil layers.

- Higher biodiversity, particularly birds, butterflies and snails.

Integrated pest management, due to increased presence of birds, ants and fungi, as well as improved livestock resistance from improved nutrition. 51

Compassion in World Farming, Case Study CIF Silvopastoral Systems."

• Solutions to Growing Foods Without Soil

- **Vertical Farming,** a way of growing vegetables without soil. This is a technical type of farming using robotic arms and humans. This type of farming is especially useful in areas where fertile land for growing greens is not available. An example of this type of farming exists at a farm called the Plenty Tigris Farm in South San Francisco, CA. Seeds are automatically planted in a potting medium consisting of scentless shreds of coconut husk, peat moss and perlite that are placed on a conveyor belt to incubate in a warm incubation room that is hidden under heavy black tarps for 14 days. One has to wear black sunglasses to protect the eyes from the intense white light. Then, the plants are zipped out of the tarp to a staging area and planted by a robotic arm into tall skinny towers 7 to 13 feet tall, holding 40 to 150 plants. Another robot type arm places the finished towers onto an overhead zipline that moves the plants into a grow room for their final stage. After the light-purply type lights have thoroughly warmed its crop, flawless leafy greens come forth from the tiny openings.

According to Dr. Emeran Mayer, a brain-gut microbiome interaction clinician and researcher, there may not be enough polyphenols (plant compounds that boost digestion and brain health) in the vegetables even though there would be sufficient fiber with this type of farming. Besides that, it takes an excessive amount of energy to run the grow lights and there is a nutritional input concern. 52

- **Hydroponics** – Another approach to growing food without soil is using water and adding nutrients. Gaining insight from Ohio State University's Horticulture and Crop Science Lab, there are 3 most common hydroponic systems and techniques. It is necessary to consult with a professional to do the engineering with setting up this type of food growing before undertaking this project. The thee systems/technique are as follows:

- A nutrient film technique uses channels, like slots, in a long gutter. Plants sit on top of the gutter and the water runs underneath, delivering water and fertilizers throughout the day.

- A deep-water culture system is essentially a pond filled with a nutrient solution. Plants sit in rafts with the roots sliding through slots, allowing them to absorb water and nutrients in the pond when needed.

- A soilless substrate culture system is a bag or container filled with aggregate mix (rockwool, peat or coco coir) to which water and fertilizers are delivered as needed.
  "Juicy Tomatoes in Winter? Thank Hydroponics."
  53
  There are many books on Hydroponics from beginners to advanced One recommendation is: Hydroponics Unearthed by Oscar Stephens

- Another Solution for growing food without using soil is found in Chapter 16, Part II. **Aquaponics,** Growing Food Without Soil

Further Reading to Explore Food and Farming:
Farming While Black, Leah Penniman
The Jungle Effect, Daphne Miller
Omnivore's Dilemma, Michael Pollan,
Civileats.com

Let's not forget that any of these actions will also help:
- the winged ones
- the crawlers
- the swimmers
- the hoofed and claw-toed ones
- the microscopic ones

What are your comments about Food and farming?

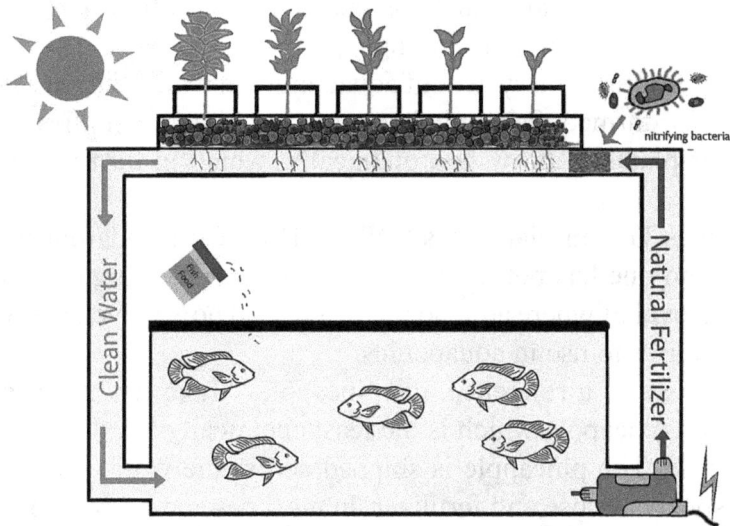

Aquaponics for Beginner, Teachers & Kids.

Chapter 16

Part II
Food & Farming

Aquaponics

As we have read in Part 1, Food and Farming, healthy soil can be critical to growing healthy foods. Through the mirror of supplying food to feed people putting less strain on the climate, let's investigate the workings of an aquaponic system to grow food without soil and even in a drought season.

To explore this venture, I have written this chapter based solely on personal interviews with Both Company's (Both Co) Board of Directors: Justin Wright, James Galvin, & Janna Ratzlaff. They created this company as graduate students at Monterey Institute of International Studies in 2014 and have moved from Both Co to pursue other interests in farming. 1

Aquaponics is a system of aquaculture in which the waste produced by farmed fish or other aquatic animals supplies nutrients for plants to grow hydroponically, which in turn purify the water. 2

James Galvin elaborates "Where Both Co stands out is that maybe no one has put in more effort to understanding a full ecological level of where aquaponics stands and how recyclable waste is a resource to use in aquaponics."

He uses as a reference, the pineapple. "Take for instance the core of a pineapple which is the least chemically filled part of the pineapple. The pineapple is shipped across the world with super intensive water use and fertilizer. In most cases we throw the core of the pineapple away but it doesn't have to be. It can be composted rather than thrown in the garbage that goes to the landfill that creates methane. Methane by law is 25 times a more lethal greenhouse gas. 49 to 120 times more potent and making our land uninhabitable. So, compost food, newspapers, paper waste (not plastic) and lose nothing as it is a nutrient. Thus, converting compost into aquaponics. If there is more than we need, it can be put into bio-digester to turn into methane and run generators in our home systems."

Galvin further explained how food is grown:

"The chain of aquaponics starts with fish. You have to feed them marine protein. If you buy it, it is usually bi-catch from unsustainable fishing. But say we're going to give them worms and minnows raised on-site. Instead:

Both Co.
- turns compost to worms
- turns worms to minnows by feeding the worms to minnows
- then fish eat the minnows

All is done with what's on sight and thus fish can grow pond veggies on top of the fish ponds.

And once the water is cycled through the first few months, it stays pristine, because the system keeps filtering it all the time and might stay cleaner than one's pool. What happens is:

There are bacteria in the air everywhere and when the fish breathe out the ammonia, the bacteria are attracted to that ammonia. The bacteria called nitroso monas comes into the pond and convert the ammonia using a metabolic process into nitrites. And those nitrites can attract other bacteria and those bacteria which are a *nitrobacterium*, family of bacteria, can convert it from nitrites to nitrates and those nitrates are consumable by the plants."

In finding out how the plants actually sit still and stay still in one place, Galvin had three major answers. "Here we go:

1) <u>Deep Water Culture</u> or DWC. They are growing in deep water. The roots are sitting in water and sometimes that's just where pond vegetation grows it, for example duckweeds. Put three in there and you have 4,000 the next month. But they are little plants and also affix nitrates. They saved the planet 50 million years ago from a similar climate problem. Another way to do deep water culture is to hold plants in place using rafts. One can use Styrofoam repurposed for a few years for food purposes; and can be housed in food-safe silicone wrap silicone around it. Can use fiberglass which is not proved food safe. Also, you can use something natural like "coco coir" or the outside stringy stuff of the coconut to make rafts. And you can plant the seeds right in that and the roots grow down and that's one way to do deep water culture.

2) <u>Nutrient Film Technique</u> or NFT is where you run a small stream of water through the bottom of the pipe and the evaporation of that water + droplets of water get warm from running through the pipe, either because you have it heated or from lights of that heat, another heat source or the sunlight. It heats the water enough that little bits of it evaporates up to the roots. And you have the plants sitting up in that little

cup of media [media refers to a growing media or what seeds are placed in to grow either granite, coco coir, sand, pebbles i.e. anything food safe and pH neutral and won't mold.

The pH is checked using an Aquarian test kit. But you would know ahead of time. If you used a rock and look at it, you know if it will impact pH as if it is a river stone. It is likely to have a big impact compared to granite which is not very inert and won't have much impact.

Another one that folks use for the grow media is a lighter weight, more expensive hydroton. These are little pebbles of clay that have air blown into them. They have 1/2-inch diameters and are a perfect sphere. You can put seeds in them and roots grow through it. Every one is the size of a marble and you put seeds in that pile of marbles. You can put pebbles of glass or anything gorgeous. For a commercial purpose, you wouldn't do that. The nutrients actually evaporate up a film in the droplets and to the roots and the roots grow down into that little river in the pipe.

3) <u>Flood Drain</u> is the third method to grow. There are several ways to do that as well. Typically, the water comes in and fills up to almost the top and then drops down to almost the bottom. You always want to have a wet zone for some moisture. You pump it up and it comes down, usually having an exit that's smaller than your entrance. And it builds up and the exit slowly drains it back down. Or you can do percolation where you run water down through it all the time through the media. The water goes back down and stays clean because any excess nutrients available in the water are all cleared out and the plants took what ever was there. And if you have healthy plants, like Both Co does, then they are looking for more. And you can choose the way you feed your system- to

keep the water a little bit filthy but not enough that the fish care or you can keep it completely pristine with no nitrates, no ammonia. The ammonia is all getting turned into nitrites and nitrates before it can build up, all turning into plants and not feeding the system enough so that there are not any nitrates left. But we like to keep it dirty and the best way to go is keep the nitrites up at maybe 40 to 80 parts per million so you always know you have a little bit more food and the plants are not out of food.

It is super-efficient and the water goes around and around."

With all this information about how aquaponics work, what else could there be? In other words, where does Both Co see aquaponics going in the future?

Janna Ratzlaff explained: "A big fat lumbering squirrel who eats the lettuce and presses me to be better, ate a cucumber yesterday. Eventually people are going to realize whether it is past the tipping point or after the tipping point that we can no longer farm with modern agriculture. Small nations or nations with less area, because they can have global impact, such as Japan are moving into aquaculture and hydroponic systems. All the farmlands are being shut down and eventually they are going to take it a step further and realize they cannot sustain the waste from these hydro or aquaponic systems and eventually there will be that movement to combine them. And I think we'll be a step further than that as we'll be moving into AEROPONICS.

Most people grow plants in this aeroponics system without any media. Say you put the seed in a little bit of coco coir and you spray a fine mist at the plant and eventually the roots grow. Usually if you are doing this at the maximum capacity, you are just giving the plants exactly what it wants. The water particles are absorbed immediately."

Galvin- "So, then you are right, you don't have to recycle. You are only sending out what the plant needs and it never has to come back (circulate). You are sending just enough water even

in a trapped contained way and the roots are growing in wet air."
Aeroponics means we don't need all that energy used in box scale
agriculture either because we can just use these expensive com-
pressors that we can get to put out 20 microns water droplets,
(that's small). You only shoot up just enough as there are sensors
that tell you the humidity that the plants like so that all the water,
the only water that needs to be brought up using any energy is just
the amount the plants want and no more. And that's possible."

Ratzlaff- "And you can go even further than that. When you are
growing in aeroponics, there's all these other crops you can add to
what you are growing. You could grow potatoes with aquaponics
but they wouldn't grow as well because there will be a lot of media getting in the way. There has been a huge success in growing
potatoes in aeroponics. Further than that, people have gone and
done multiple studies with ginseng and other crops which have
been unsustainably harvested, destroying the countryside in many
places. You can grow those in aeroponics but it is not as popular
a system."

Galvin- "A lot of the popular ultra-health foods that are super-
nutritious are grown with aeroponics."

Moreover, addressing aquaponics concerns led me to ask this
question:

Is it dangerous to eat fish when there's too much waste in the
water?

Galvin responded: "In commercial aquaponics when they
have a super-density of fish with the maximum temperature possible, the fish are mostly hungry and grow the fastest. But that also
means the fish get stomach cancer because they are stressed out.
Once a year, they have to kill their entire population and turn it
into dog food and cat food. This is more profitable for them. By
reducing the density by slivers of a penny, Both Co does aquaponics with less density. This removes the risk of environmental damage as well. For instance, some commercial businesses with super-
density in super-filthy tanks completely flush every 6 months and
destroys a complete watershed."

In addition to the how aquaponics work and its concern, there
is the financial prospective. Justin Wright, the financial backer and

CEO, made an interesting comment because of this once-a-year flush out of commercial systems. He pointed out that "if you are a company where there is a public offering that anyone can have stock in your company, a stock holder can sue you if you are not trying to maximize your profit to the most that you can. Therefore, Both Co is not looking at this direction even if they choose someday to operate on a commercial level."

Although Wright and fellow members of the board meet weekly for ideas and operational interests, they are constantly considering beyond small systems, with raising and selling culinary herbs at Farmer's Markets and personally eating the fish. Consulting, creating auto cads, kits to sell, different pods in various places and a warehouse with a large system (aquaponics and aeroponics) are matters of their discussion. Both Co wants to teach folks to grow more sustainably. Wright further communicated that "Both Co has to find investors who have a view like ours, rather than bankers. Both Co is looking at becoming a B-Corporation (B-Corp)" described as a new type of company that uses the power of business to solve social and environmental problems, often known as a business that inspires people, not just profit." [3]

"Both Co is trying to establish a business whose framework functions so they can't get sued if they don't maximize profit, while also having non-stressed out fish and workers that are getting paid well. Both Co is working with the B-lab to hopefully design the business or a part of the business to be involved in social entrepreneurship."

Galvin noted that "there are 31 states that acknowledge the benefit election so when you set up a corporation; C-corps, S-corps or an LLC, you can elect to be a Benefit Corp (B-Corp). That means a certain bracket for your taxes and it can help to acknowledge that you can be sustainable and not try to maximize your profit. You can go to B-lab and say I elected to be a Benefit corporation and my state acknowledges benefit elections. And will you help me to work through the process of becoming B-certified? And that is where you become a B-corporation, not just a Benefit corp. Then Both Co, our non-profit will hopefully own one part of that corporation so that we operate both. We could have a big warehouse where we grow sustainably and we can also have a small sustainable system, one that brings students outdoors. In a smaller pond, perhaps near the commercial warehouse, the veggies can't be sold as we bring in school kids for an education component- field trips where it is not a bi-hazard to allow visitors."

Ratzlaff suggested that "Both Co can be involved in restoration projects locally and globally, having different communities included while teaching them to have a more sustainable livelihood.

As with any project requiring the use of water, it is of essence to use water that is considered clean, not contaminated."

ACTION

THE NEW STORY OF HEALING & SOLUTIONS
Every solution and resource births a new story for change. If we want to see changes, we must take action.

In this part of the chapter are possible solutions and resources to help inspire actions so we can have opportunities to live new or revived stories to:

A) Awaken to aquaponics and acropontics help reduce the greenhouse gases in the atmosphere, benefitting both people and planet.

B) Find any action item as a pathway towards a climate career, social justice career, research or volunteer project that is just waiting for you to choose from. That is, unless you have already been triggered into action.

List of possible action items to pursue:

Resources for careers or projects with Aquaponics
Search: INDEED- aquaponics job finder [4]
Search: Career Opportunities in Aquaponics. 5

Additional Resources to read about aquaponics:
Search backyard aquaponics 6
Search the aquaponics source 7
Search aquaponics solutions 8

Suggested reading:
Sylvia Bernstein, Aquaponic Gardening: A Step-By-Step
    Guide to Raising Vegetables and Fish Together 9

Let's not forget that any of these actions will also help:
- the winged ones
- the crawlers
- the swimmers
- the hoofed and claw-toed ones
- the microscopic ones

List 3 things you learned about aquaponics?

Types of Biomass

Chapter 17

Biomass As Renewable Energy

We have another renewable source of energy, right under our feet- wood, crops, manure, and some garbage.

Wikipedia says that Biomass is plant or animal material used for energy production, or in various industrial processes as raw substance, for a range of products. It can be purposely grown energy crops, wood or forest residues, waste from food crops, horticulture, food processing, animal farming, or human waste from sewage plants." [1]

Biomass is definitely a renewable energy source because we can always grow more trees and crops, and waste will always exist. When burned, the chemical energy in biomass is released as heat or converted to liquid biofuels or biogas that can be burned as fuels.

In other words, in a direct combustion system, biomass is burned in a combustor or furnace to generate hot gas, which is fed into a boiler to generate steam, which is expanded through a steam turbine or steam engine to produce mechanical or electrical energy. [2]

As far as uses go, biomass energy supports U.S. agricultural and forest-product industries. The main biomass feedstocks for power are paper mill residue, lumber mill scrap, and municipal waste. For biomass fuels, the most common feedstocks used today are corn grain (for ethanol) and soybeans (for biodiesel). 3

Biomass is thought of as being carbon neutral since the CO_2 being released is the same amount as was removed from the atmosphere during the plant's lifetime. We have to use it sparingly and consider however, that growing enough biofuels (switchgrass, sugar cane, corn) to replace all the fossil fuels that we may rely upon, would use up all the farmland that we now use to grow food, or require massive deforestation. Consider also that the machinery and fertilizers, if used, would also generate greenhouse gases.

Now that we have touched on Biomass materials, I want to end this chapter in a different way. Ground yourself in nature. I want you to put your hands in nature. Go out and collect biomass materials: leaves, twigs, pine cones, bark, leaves, acorns, seeds, dried flowers, small rocks, acorns and a variety of other types of material that nature offers. Get outdoors and free from computers, mobile phones and other technical gadgets and get into a creative stage. Create alone, with your children, grandchildren, classroom students or anyone that wants to join in. A website is provided at the end of the action section to give you craft and activity ideas.

ACTION

THE NEW STORY OF HEALING & SOLUTIONS

Every solution and resource births a new story for change. If we want to see changes, we must take action.

In this part of the chapter are possible solutions and resources to help inspire actions so we can have opportunities to live new or revived stories to:

    A) Awaken to the use of biomass and biofuels as a source of renewable energy to help reduce the greenhouse gases in the atmosphere, benefitting both people and planet.

B) Find any action item as a pathway towards a climate career, social justice career, research or volunteer project that is just waiting for you to choose from. That is, unless you have already been triggered nto action.

Possible action items to pursue

■ **Learn more about Biomass and Careers in Biofuels** Search: U.S. Bureau of Labor Statistics-Careers in Biofuels. 4

■ Search: Office of Energy Efficiency & Renewable Energy-STEM and Education. 5

■ Search: GE Renewable Energy Careers. 6

Jobs at GE. 7

Have some fun with crafts and activities using items from nature:
https://www.weareteachers.com/nature-crafts/ 8

Let's not forget that any of these actions will also help:
• the winged ones
• the crawlers
• the swimmers
• the hoofed and claw-toed ones
• the microscopic ones

The Elemental Mirror of Fire

What creates excessive fires?

**What's good or bad about Fire that nourishes
or perishes?**

What's justifiable about a vast number of wildfires?

**What goes on with Fire disasters that is justifiable
for struggling people or people-of-color in our world?**

Here's what the late West African writer and teacher, Malidoma Patrice Some, once said about Fire: "In the indigenous mind, fire kindles and sustains an animating and pervasive energy in all that lives. It is in the water that runs. It is in the trees, the rocks, the earth and in ourselves. It is the mediator between worlds since it is very close to the purest form of energy. Fire is the rising force that makes us do, see, feel, love and hate. Fire has great power, both outside us and within us.

Even if we don't know all the answers right now, let's open up to some of the issues and solutions presented in chapters 18-19 of this section. Please commit to peacefully act on at least one or some of them to help create a better world.

Air tanker releasing fire retardant over the Big Sur Soberanes Fire.

Chapter 18

Fires

Right off the top of the bat, we know what fires can do to the climate and persons inhabiting the planet. And yes, greenhouse gases are emitted. According to a World Forum Economic article, "While emissions from fires contribute to the warming, they are far from the only source of carbon dioxide emissions. Burning fossil fuels, for example to power our homes and our cars, is the leading source of emissions globally." Some may agree or disagree. We do know that some houses are being built in the wildland urban interface where there are forests and other natural vegetation. This can lead to more wildfires often caused by humans resulting in greater losses to human lives and property. In other words, a tiny spark has the potential to spread into a raging fire, fueled by higher temperatures, drier conditions and rising emissions. 1

Without condoning wildfires as a wonderful thing, there is the reality that if they occur, there are some benefits from them.

It is not uncommon to see public literature about fires frequently posted for State Park patrons. One such information piece I spotted while hiking in Pfeiffer Big Sur State Park, Big Sur, CA, reads as follows:

"This canyon has been affected by fire and debris flow from two major fires. The Basin Complex Fire burned 162,818 acres in 2008, and the Soberanes Fire burned 132,127 acres in 2016. Fire is a natural disturbance with many benefits to the ecosystem. It burns debris and dead vegetation, recycling nutrients into the soil, and improves wildfire habitat. Clearing the understory allows more sunlight to reach the forest floor and creates space for new plants to grow. Many plants require fire for seed germination, and some conifers have cones that require heat from fire to open. Heavy rains after a fire can cause mudflows, which transport sediment and debris to new areas.

Humans have suppressed fires, altering their frequency and intensity in California, but regular wildfires actually reduce the risk of larger, more intense fires. Many plants, such as coast redwoods, regrow immediately after fires. You can see the regrowth as you hike the trails affected by the fires."

I also had personal experiences with both the Basin Complex Fire and the Soberanes Fire, and was affected intensely by the Soberanes Fire as it was a close to my home and I chose to not evacuate. Rather, I stayed with my husband and son to do even more fire clearance that was already in place. This was a first-hand, everyday emotional and physical account whereas, during the Basin Complex Fire, I evacuated.

The Soberanes Fire raged the Big Sur coast in the summer of 2016. It started on July 22 and was not extinguished completely until the fall of that year.

This event was a direct result of drought caused by global

warmth with the element of human carelessness or either an illegal campfire on the Soberanes scenic trail. There is a great risk when one leaves behind smoldering embers or fire coals that are not fully extinguished or even tosses a cigarette butt in any drought-prone environment. All it takes are these elements - heat, fuel and oxygen and a fire starts.

It was startling to know the actual final incident public information of the Soberanes CalFire report of October 13, 2016 when the fires were 100% contained. The fire burned 132,127 acres (94,933 acres of Los Padres National Forest) that included the locations of Soberanes Creek, Garrapata State Park, Palo Colorado and north of Big Sur. Although devastating, alarming and sad, the firefighters were top notch and were the best strategists in analyzing the Soberanes fire. The Fire Analyst and Foresters communicated this very well to the community. 2

One can learn a lot about the career path for a Fire Behavior Analyst as well as a Forester by taking into consideration all the responses given by one. I interviewed Dennis Burns, Fire Behavior Analyst for the Soberanes Fire while the fire was still burning. He definitely had an inside seat to understanding what it takes to help cool the planet when there is a wildland or forest fire. Here are the responses:

Interview Questions:

1) *Would you say a fire, the Soberanes Fire for instance, would have looked the same 10 years ago? 20 years ago?*

Both 10 and 20 years ago the Soberanes Fire would have been much worse. The fire started in a fuel regime that had not seen fire in 100 years which is why there was explosive growth on this fire for the first several days. The fire growth has now slowed as it is in the footprint of the Basin Fire. California chaparral generally reaches maturity in 20 to 25 years at which point it will burn at its most extreme. Given that the Basin Fire was 8 years ago (since this interview), the fuel bed is not nearly as deep and continuous as such. The fire has slowed and the effects on the landscape are not as damaging.

2) *Would you say that our California climate is warmer now than 10 years ago? 20 years ago?*

I think the climatology speaks for itself. California is warmer now than it was 10 years ago. Of a bigger concern is the persistent drought we have been in. We are seeing our normal fire season mid-June to mid-October extended to the point that we are responding to vegetation fires almost year-round.

3) *What year did you notice an increase in wildfires? And if so, where was this?*

There really has been no significant increase in the number of wildfires in California. We see year to year fluctuations depending on whether or not we get lightning storms over the state. For example, right now we are running about average for the number of wildfires. Last year we were below average. What we have seen is the number of acres burned which has been increasing over the past ten years.

4) *Are forest fires more frequent now and spreading further? If so, why?*

Fires are not more frequent however our fire season is lasting longer (see above). What has become more frequent are large fires (10,000 acres or more). As to the reason why I see three factors; more people are living in the wildland urban interface (WUI). With fires in the WUI, firefighters are having to focus suppression efforts on saving lives and homes rather than perimeter control which allows some fires to get bigger. The second is poor forest management practices, whether you agree with logging or not when the forests were being logged, they were managed much better. Fuels were not accumulating to the degree that they are now. More fuel equals bigger fires. The last is we now have more areas of our national forests being declared as wilderness areas. As such we are limited on the types of resources that we are allowed to use to suppress wildfires. Further, there is no standard from one wilderness area to the next. Some are not allowed to even use saws where in others, it ok.

5) *What effects does the precipitation levels and soil moisture have in relation to forest fires, whether caused by lightning or human error?*

Precipitation is critical to the spread and size of wildfires. The reason for this is what is referred to as live fuel moisture, with less rainfall the live vegetation has less water in it. In a more typical or what used to be normal, most living vegetation will be at its lowest fuel moisture in the late fall (when we see the leaves drop). This would be the time that the live fuels would be the most susceptible to burn. It is not uncommon now to see the live fuel moistures at late season levels as early as mid-July or sooner. The result is not only the dead fuels burning but the live fuels are also burning.

Questions 6-10 are answered together:

6) *Knowing all that a Fire Analyst should know to help predict fire patterns, would you recommend to a youth to pursue this career path? And if so, are there any pros or cons?*

7) *How long have you been a Fire Analyst?*

8) *Were there other fields or careers that lead to your path as a Fire Analyst?*

9) *What are the qualifications, the course of study or degree required to become a fire analyst?*

10) *How long have you been in this field?*

I think I can answer questions 6-10 altogether. I have been a fire fighter 39 years (as of this 2016 interview). My present job is an engine captain for the Livermore Pleasanton Fire Dept. I have been a Fire Behavior Analyst for 20 years. I also manage our department's prescribed burning program. This program is where we work with the park district to use fire to get rid of non-native species and reestablish native grasses and wild flowers.

I have a BS degree in Fire Management from California State University, Sacramento which is a special major program. Any degree in biology, forestry or environmental science would also be a path to follow. To become a Fire Behavior Analyst (FBAN) requires an extra 36 semester units of specific fire behavior training that is administered by the National Wildfire Coordination Group (NWCG). With the class room training done you become a trainee and work with a fully qualified FBAN until such time as that person feels you are fully qualified.

In general, it would take about 10 years as a fire fighter before you would be ready to become a FBAN.

As a career path, I would recommend it with the caveat that it is not suited for everyone. It is physically demanding, has long days and a lot of time away from family. The pros are you work with great people all focused on the same mission. Every day brings something different and you get to see and work in places many Americans never see. The biggest plus is you are outside, not in an office and you are doing something positive for the environment and mankind.

11) *What has rewarded or satisfied you to continue in this career path?*

As corny as it may sound, I have wanted to be a fire fighter as long as I can remember. I have never considered doing anything else. To me, I have won the lottery. The rewards and satisfaction are simply; I interact with people on what is usually the worst day of their lives and help to make it a little better.

A final note from Dennis Burns: One thing I know for sure is that our drought is tied to the El Nino/La Nina cycle. What if any role does global warming plays in that? I can't say for sure but it seems logical to me that the increase in global temperatures has to have a role in this. Also, I can clarify as far as the subject of logging, I am not an advocate of clear-cut practices that were done in some areas in the past but a more managed thinning. If you look at photos of our forests from the turn of the century, you would see open stands, often park-like.

Now our forests are becoming choked with growth. This means there is more fuel for fires. It is easier for fire to get into the canopy. Thus, crown fire as well as the trees become more susceptible to the spread of disease as we are seeing in the Sierras (California) right now. 3

As concerns regional fires, there have been many that have raged throughout the United States. California's recent pattern of building residential areas in the wildland urban interface exposes people and property to wildfire sometimes with deadly results. These same patterns degrade adjacent wild habitats through exposure to human influences, such as fragmentation, changes in runoff, light and chemical pollution, impacts of pets and trampling.

Sadly, the most critical and deadly fire was a small urban community of houses in the heart of a wildland urban interface that took lots of human lives. It was situated in California called the Camp Fire in Paradise, CA. It started November 9, 2018. This deadly fire burned 85 people to death according to a CalFire report. It was caused by a PGE utility company faulty power line and fueled by homes and forests from an upsweep of gusty winds. An estimated 153,336 acres burned and wiped out more than 13,600 homes. It was the most destructive wildfire in California history and the nation's deadliest in a century. 4

On the other hand, moist forested areas are most likely to become an issue with wildfires as conditions become parched and arid with the rise in temperature. Consider the moist forested area called the "rainforests" that suffer when there is a wildfire. Even if trees do not burn, they can die from drought.

Since trees pump water into the air through their leaves, then fewer trees mean less rain. As the trees burn and decay, all the carbon in their timber will turn to carbon dioxide, which raises temperatures still further.

Looking through a social justice lens, "A general perception is that communities most affected by wildfires are affluent people living in rural and suburban communities near forested areas," said Ian Davies, lead author and graduate student in the University of Washington School of Environmental and Forest Services. But there are actually millions of people who live in areas that have a high wildfire potential and are very poor, or don't have access to vehicles or other resources, which makes it difficult to adapt or recover from a wildfire disaster."

Another way to call this is social injustice due to the imbalance of resources as people-of-color communities experience greater vulnerability to wildfires and cannot recover so easily. Of course, some socioeconomic factors come into play such as how much money is saved up, if below the poverty line, not to mention if elderly and disabled.

Researchers with the University of Washington Environmental and Forest Sciences indicate that Native Americans are six times more likely than others to live in areas most prone to wildfires. For Native Americans have been historically forced to relocate onto reservations while confronting socioeconomic obstacles. 5

.

ACTION

THE NEW STORY OF HEALING & SOLUTIONS

Every solution and resource births a new story for change. If we want to see changes, we must take action.

In this part of the chapter are possible solutions and resources to help inspire actions so we can have opportunities to live new or revived stories to:

> A) Solve the contradiction of lessening the increase of wildfires and still safely get rid of the excessive forest fuels without starting fires that increases greenhouse gases in the atmosphere

B) Find any action item as a pathway towards a climate career, social justice career, research or volunteer project that is just waiting for you to choose from. That is, unless you have already been triggered into action.

Possible Solutions to Reduce Greenhouse Gases & Prevent Wildfires

1. **PLANT TREES**- Let's just plant more trees for the future where trees have been burned. There is a need to plant as many trees as possible in forests that have burned out. Let's be thankful for those persons and organizations that are helping to mitigate the climate by doing this.

Refer to **Future for the Trees** resource that is also a solution for deforestation noted in Chapter 8.
Search for Trees for the Future 6 & the Facebook page. 7

2. **Contribute to ONE TREE PLANTED** - One tree planted is a 501(c)(3) nonprofit on a mission to make it simple for anyone to help the environment by planting trees. Their projects span the globe and are done in partnership with local communities and knowledgeable experts to create an impact for nature, people, and wildlife. Restoration helps to rebuild forests after fires and floods, provide jobs for social impact, and restore biodiversity. Many projects have overlapping objectives, creating a combination of benefits that contribute to the UN's Sustainable Development Goals. Learn more at onetreeplanted.org.
https://onetreeplanted.org/ 8
hello@onetreeplanted.org   Phone: 800-408-7850
145 Pine Haven Shores Rd #1000D
Shelburne, Vermont, 05482, USA

3. **Partnering with Indigenous Communities**
"Increasingly, indigenous peoples are recognized for their ability to contribute to solutions for our mounting fire challenges. By providing a supportive framework, the U.S.-based Indigenous Peoples Burning Network is uniquely positioned to elevate tribal contributions in this shared journey." Search Indigenous Peoples Burning Network. 9

"After many years of Australia's northern savannas being ravaged by huge wildfires, emitting large volumes of greenhouse gases, Australia is gradually returning land management of the region back to its original human inhabitants. The traditional owners have always known that lighting relatively small fires in cooler months can prevent large wildfires in summer (and therefore reduce net carbon emissions). In Working for Australia's Indigenous Communities we're combining traditional ecological knowledge with the latest in fire science to benefit vast areas of Outback Australia." Search for Working for Australia's Indigenous Communities. 10

4. **Explore: Conservation Biology Institute**
CBI is partnering with The Nature Conservancy to explore community design principles to provide wildfire risk reduction benefits. The team is using GIS to compare former land use patterns to innovative design measures that include the location of irrigated greenspaces between structures and wildlands to reduce wildfire risks. Greenspaces, which can be parkland, orchards, and other irrigated land uses, can act as ember-catchers and provide strategic areas for escape from fire or for fire-fighters to stage their battles against future wildland fires. Search for job and career opportunities at Conservation Biology Institute or

Contact:
Deanne DiPietro, Senior Sciences
Coordinator: 707-477-6515
136 SW Washington Avenue, Suite 202
Corvallis, OR 97333
ph. (541) 757-0687

Careers:
Search Conservation Biology Institute jobs. 11

Get Different Perspectives by searching for yourself:

5. **Prescribed Burning** is one suggested solution to
   reducing wildfires according to Dr. Paul Hessburg,
   Research Ecologist with Pacific Northwest Research
   Station, US Forest Service. He has studied historical
   and modern era forests of the Inland West for over 32
   years. We can use prescribed burning to intentionally thin
   out trees and burn up dead fuels. It's going to create al
   ready-burned patches on the landscape that will resist the
   flow of future fires. Prescribed burning produces so much
   less smoke than wildfires do. There is an issue and that is
   prescribed burning smoke is currently regulated under air
   quality rules as an avoidable nuisance.

   **There are additional perspectives available about
   prescribed burning from further research.**

6. **Managed Wildfires.** Another belief of Dr. Hessburg.
   He suggests that instead of putting all the fires out, we
   need to put some of them back to work thinning forests
   and reducing dead fuels. This can be done by herding
   wildfires around the landscape when it's appropriate to do
   so. Hessburg considers it a social problem because both of
   his solutions to reduce wildfires are controversial.
   Search for Dr. Paul Hessburg's TED talk on "Why
   Wildfires Have Gotten Worse and what we can do about

it" or Dr. Paul Hessburg, Senior Research Scientist, US Forest Service. 12

Let's not forget that any of these actions also help:
- the winged ones
- the crawlers
- the swimmers
- the hoofed and claw-toed ones
- the microscopic ones

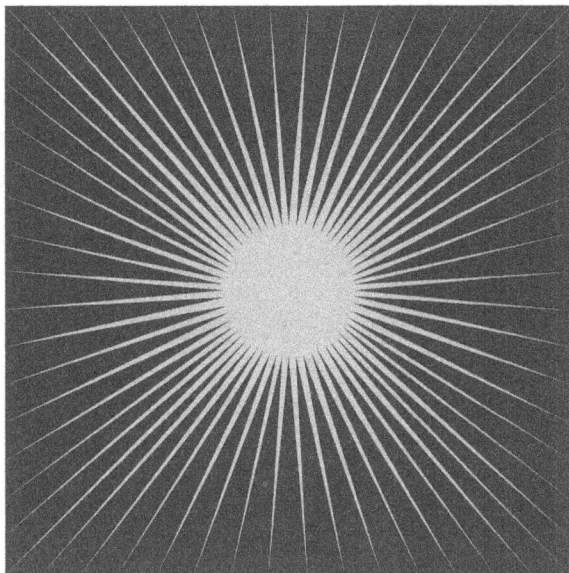

Sun

Chapter 19

Solar As Renewable Energy

We look at the power of the Sun, our natural heat source, and how that source can benefit the planet and its people. What can be said about this energy source?

Scientific World reports, "Solar Energy is one of the most abundant renewable energy sources in the world." It can be used for many things by producing chemical reactions, heat and generating electricity.

Solar energy is a clean, non-polluting source. If properly utilized, it will meet a large proportion of the world's energy needs in the future.

Solar energy works by absorbing solar light from sunlight - most of the light spectrum, plus half of the ultraviolet and infrared radiation - and converting it into electrical energy for use in homes or workplaces.

Here is what I wrote about this energy source:

SOLAR POWER

Wonderful, majestic Powerful Sun
At the dawning of day, here you come
Your huge glorious ball of energy
Combining with Earth's synergy

Your wonderful beams shine on me
Turning what you want to electricity
With solar power to charge batteries
And your image hang in many galleries

Such a BIG ball of hydrogen
Creating all that heat for the world's acumen
Held together with a lot of gravity
With compound pressure in your cavity

Influencing all the planets, moons, asteroids and comets in our
solar system
Your atoms collide bringing forth a place to listen
For your forcefulness to blend and make helium
Lightly lifting balloons and airships, such premium

You release your energy promoting a chain reaction
Your central core merging to nuclear fusion compassion
You transfer your heat or convection with glorious rays that rover
What powerful build-up over and over
With such a hot core of 27 million degrees

Traveling through the container of stars or photosphere to seize
The shining that is ever so bright
Emitting heat, charged particles and light

That chemical reaction from its powerful heat
Making life possible on Earth, like a BIG heart beat
The heat allows gas and liquids on many planets and moons
Even creating icy comets forming a disk or halo, all so soon

The particles of bright
Giving and continually emitting even more light
Calling itself a solar wind
Between the stars of interstellar-traveling billions of miles away,
our FRIEND.

©LaVerne McLeod, March 2022

ACTION

THE NEW STORY OF HEALING & SOLUTIONS

Every solution and resource births a new story for change. If we want to see changes, we must take action.

In this part of the chapter are possible solutions and resources to help inspire actions so we can have opportunities to live new or revived stories to:

A) Awaken to the consequences that solar power is a great source of renewable energy that will reduce the green house gases in the atmosphere, benefitting both people and planet.

B) Find any action item as a pathway towards a climate career, social justice career, research or volunteer project that is just waiting for you to choose from. That is, unless you have already been triggered into action.

Let us embrace the sun's power as it flows into our vehicles, homes, schools, businesses or wherever needed.

Some Resources You Might Explore:

Search -US Department of Energy: Planning a Home Solar Electric System. 1

Search-Environment America Research & Policy Center: Solar on Superstores 2

Search-Use Solar Energy to charge Electric Vehicles: EV100 Member Page 3

Search -Solar Energy Jobs 4

Search -Solar Energy International 5

Let's not forget that any of these actions will also help:
- the winged ones
- the crawlers
- the swimmers
- the hoofed and claw-toed ones
- the microscopic ones

Transportation-City Infrastructure

Chapter 20

Preventative Infrastructure

"Buildings in the U.S. account for around 40% of the country's total emissions, therefore it is vitally important that we prioritize energy efficiency and conservation in construction + building operations."

The word "infrastructure" means "infra" below and "structure" edifice, building or construction. Oxford Dictionary defines it as "the basic physical and organizational structures and facilities (e.g. buildings, roads, power supplies) needed for the operation of a society or enterprise." 1

Therefore, this all-encompassing word called "infrastructure" covers a lot of territory that equates to "support." This would include the highway underpasses where some homeless folks sleep, the streets we walk on, the public green spaces (vegetated land or water within an urban area) where community members can gather, including our home. If you live in an urban area, wouldn't you want to have a place to physically exercise run and play; socialize; a place where the immune system is improved by exposure to beneficial microbiota and a place that decreases noise and air pollution? Green spaces could be the default for the way we design cities, build buildings, and configure transportation.

Transportation and city planning are also key components in green infrastructure. Although, for the people that rely on public transportation, who dwell in urban areas, are often left out of the conversations around green infrastructure. In order to build thriving cities, green infrastructure is to be looked at from the whole rather than just a part of the city that accommodates the rich. Thus, inclusivity and equity for everyone's convenience and comfort are vital factors to bring to planning meetings that intersect with climate and social justice.

Of equal importance is redesigning homes and city infrastructure to deal with preventing excessive damage that could be caused by extremely warming temperatures. This disparity shows up in some inner city BIPOC neighborhoods.

According to Climate Central, that researches and reports climate change impacts:

"Paved roads, parking lots, and buildings absorb and retain heat during the day and radiate that heat back into the surrounding air. Neighborhoods in a highly-developed city can experience mid-afternoon temperatures that are 15°F to 20°F hotter than outlying areas with more vegetation and less development." 2

Here is what the United States Environmental Protection Agency has to say about heat equity:

"Heat equity refers to the development of policies and practices that mitigate heat islands and help people adapt to the impacts of extreme heat in a way that reduces the inequitable distribution of risks across different populations within the same urban area. Heat equity can be improved by ensuring that all residents have equal access to local heat island mitigation programs and tailoring a city's response to severe heat to meet the unique needs of the

most at-risk residents. Some local governments have found that working to reduce their heat islands in this way can help at-risk residents and create more livable cities for everyone." 3

Climate Central also implies that BIPOC communities are disproportionately affected by urban heat islands (UHI). This means metropolitan areas that are hotter than their outlying regions, with the impacts felt most during summer months. 4

The consequences of redlining where white affluent communities and BIPOC lower income communities are separated fuels the fiery situation. Local and federal officials in the 20th century, enacted policies that reinforced racial segregation in cities and diverted investment away from minority neighborhoods in ways that created large disparities in the urban heat environment.

Take for example in Richmond, VA, the New York Times exposed a story on a hot summer day in the Gilpin neighborhood that quickly became one of the muggiest, suffocating, hot part of Richmond, like an oven. And beyond the climbing climate-related temperatures there are other reasons that the heat rises. 5

There are few trees along the sidewalks to shield people from the sun's extremely hot rays. The low-income public housing lacks central air conditioning and the front yards are paved with concrete which absorbs and traps heat. The ZIP code has among the highest rates of heat-related ambulance calls in the city. This is also common in other cities. To list a few this includes Baltimore, Dallas, Denver, Miami, Portland and New York. The temperature variance can be 5 to 20 degrees Fahrenheit hotter during the summer than in wealthier, whiter parts of the same city. 6

For an interactive map that shows areas that are threatened by rising temperatures, search for Climate Central Coastal Map Screening Tool or go to Coastal Risk Screening Tool. 7

Preventative infrastructure or the basic physical and organizational structures and facilities involves planning for climate disasters. It is imperative to design or re-design cities that can be damaged by storms and_floods. Generally speaking, the focus to address is wherever irregular intervals of rising waters exist, such as sea levels or overflowing rivers and waterways.

Climate Central's 2020 study revealed that United States affordable housing exposed to flood risks is projected to triple by 2050. This would include New York City (4,774 units), Atlantic City (3,167 units) and Boston (3,042 units). This means there is a lot of work to be done. 8

Preventative Infrastructure also means doing something about flooding. A study from the Nature Climate Change journal found that the annual cost of flooding across the U.S. will hit $40 billion annually by 2050, compared with $32 billion currently. The study said that "while today it is mainly white, poor constituencies who are in the firing line, in the future, predominantly Black communities will be the worst hit. Black communities will be disproportionately saddled with billions of dollars of losses because of climate change as flooding risks grow in the coming decades." 9

For an interactive map that shows areas that are threatened by sea level rise and coastal flooding, search for Coastal Risk Screening Tool -Climate Central coastal map or go to and select Affordable Housing. 10

ACTION

THE NEW STORY OF HEALING & SOLUTIONS

Every solution and resource births a new story for change. If we want to see changes, we must take action.

In this part of the chapter are possible solutions and resources to help inspire actions so we can have opportunities to live new or revived stories to:

A) Awaken to the issues and find solutions to improve infrastructure that will be beneficial to ALL people no matter what race or other differences

B) Find any action item as a pathway towards a climate career, social justice career, research or volunteer project that is just waiting for you to choose from. That is, unless you have already been triggered into action.

- **Things to Do to Help Mitigate Heat Islands**

  - Trees and Vegetation - Increasing tree and vegetation cover lowers surface and air temperatures by providing shade and cooling through evapotranspiration. Trees and vegetation can also reduce storm water runoff and protect against erosion.

  - Green Roofs - Growing a vegetative layer (plants, shrubs, grasses, and/or trees) on a rooftop reduces temperatures of the roof surface and the surrounding air and improves storm water management. Also called "rooftop gardens" or "eco-roofs," green roofs achieve these benefits by providing shade and removing heat from the air through evapotranspiration.

  - Cool Roofs - Installing a cool roof – one made of materials or coatings that significantly reflect sunlight and heat away from a building – reduces roof temperatures, increases the comfort of occupants, and lowers energy demand.

  - Cool Pavements - Using paving materials on sidewalks, parking lots, and streets that remain cooler than conventional pavements (by reflecting more solar energy and enhancing water evaporation) not only cools the pavement surface and surrounding air, but can also reduce storm water runoff and improve nighttime visibility.

  - Smart Growth - These practices cover a range of development and conservation strategies that help protect the natural environment and at the same time make our communities more attractive, economically stronger, and more livable. Search-United States Environmental Protection Agency, Heat Islands & Equity. 11

• **Learn about Adapting to Heat-** Research the following possibilities:

Cities can use the information and resources to begin developing comprehensive plans to adapt to heat.

- Comprehensive Heat Response Planning
- Forecasting and Monitoring
- Education and Awareness
- Responses to Heat Waves
- Infrastructure Improvements

Search- NIHHIS About Urban Heat Islands. 12

City Planning Suggestions

Connect with your City or County Planning Department to know their information and meeting times and get involved to create change.

• Use **green infrastructure** when planning cities and transportation. This encompasses a variety of water management practices, such as vegetated rooftops, roadside plantings, absorbent gardens, and other measures that capture, filter, and reduce storm water. In doing so, it cuts down on the amount of flooding and reduces the polluted runoff that reaches sewers, streams, rivers, lakes, and oceans. Green infrastructure captures the rain where it falls. It mimics natural hydrological processes and uses natural elements such as soil and plants to turn rainfall into a resource instead of a waste. It also increases the quality and quantity of local water supplies and provides myriad other environmental, economic, and health benefits.

• Plan cities using green infrastructure and include transportation as BIPOC and poor communities depend on it. Enlist key BI-

POC community figures to go to city planners to discuss a reason for change for all people.

• Mangrove Bushes-

Another naturalistic thing we can do to protect warm region coastal areas from flooding is to use a natural barrier. Plant mangrove bushes and other coastal barrier plants. These are found in Florida. They create barriers to storm waves in the form of shoals, salt marshes. This is great for developing nations who cannot afford to build coastal defenses. Search-Mangrove Tree Roots-Mangrove Information and Mangrove Types. 13

Let's not forget that any of these actions will also help:
- the winged ones
- the crawlers
- the swimmers
- the hoofed and claw-toed ones
- the microscopic ones

What's the infrastructure like where you live?

Chapter 21

Social Justice & Climate Justice Resources

We have to work together to create a healed and nurtured planet. A planet is not healed until it is unified and inclusive with all its people. Therefore, I am including some social justice resources to enhance climate justice. The following resources are available to the public. Please use them at your own discretion.

Social Justice Resources

Consider ordering from and supporting a BIPOC *(Black, Indigenous and Persons of Color)* bookstore or your favorite bookstore

Children Books on Racism and Getting Along

- Let's talk about race - Julius Lester
- Desmond and the mean word - Desmond Tutu
- Hidden Figured: Young reader's edition - Margot Lee Shetterly
- Skin Like Mine - LaTashia M. Perry
- Hair Like Mine - LaTashia M. Perry
- A kid's book about racism - Jelani Memory
- I am enough - Grace Byers.
- Ron's Big Mission - Corrine Naden
- Anti-Racist Baby Picture Book - Ibram X. Kendi
- The Day You Begin – Jacqueline Woodson & Illus. by Rafael Lopez

Adult Books on Racism and Getting Along
- The unshackled mind: How to positively cope w/ racism and bigotry – Almas J. Sami
- How to be an antiracist – Ihram X. Kendi
- So, you want to talk about race – Ijeoma Oluo
- My grandmothers' hands: Racialized trauma & the pathway to mending our hearts & bodies – Resmaa Menakem
- The sum of us – Heather McGhee
- Lead from the outside – Stacey Abrams

Books to Teach Kids About Climate Change
Search: 14 Actually Good Books to Teach Kids About

Climate Change
- Fatima's Great Outdoors-Ambreen Tariq
- Sydney and Simon Go Green! - Paul Reynolds
- The Boy Who Harnessed the Wind-William Kamkwamba & Bryan Mealer
- Thank You Earth - April Pulley Sayre
- Moth- Isabel Thomas & Daniel Egneus
- The Lorax- Dr. Seuss

Teacher Resources & contact info
Search: Articles-Scholastic Teaching Tools

PreK-8
Search: Vanderbilt University-Teaching Race: Pedagogy and Practice

College
Search: Resources for Teaching About Race and Racism with the New York Times

Search: Self-Care: National Museum of African American History and Culture

Technology based education on Racism.

- App - Guess My Race – for older kids and adults.
- App - Who am I? – designed to instigate a dialogue about race between adults and young children.
- YouTube video read on the book "Wings" K-1 grade
- Woke Kindergarten on YouTube.

Teacher, Kid, Parent- Black History Month Resources
Search: Some other available resources
- 8 Black History Month Books and Resources for Kids -Justice Jonesie
- Celebrating Black History and Diversity Builds Self-Esteem and Empathy -Too Small to Fail
- African American History for Kids -University of Illinois Extension
- Top 15 children's books for black history month -Family Education
- 45 Books to Teach Children About Black History – The Culture

Climate Justice Resources

- Communities for a Sustainable Monterey County- (Climate Change-Teacher Resources. NEA, NASA Climate Kids, NOAA, Monterey Bay Aquarium, The Wise Owl Factory, Kiss the Ground)

- Hawken, P. (Ed.). (2017) ***Drawdown:*** *The Most Comprehensive Plan Ever Proposed to Reverse Global Warming.* Penguin Books

- Elders Climate Action

- National Food to Farm Network

- Empowering Native Communities (Navajo & Hopi)
- Further Reading to Search: **Two Distinguished Scientists on How to Rescue Humanity,** The Anthropocene

demands a massive realignment of priorities by Charles F. Kennel and Martin Rees, August 19, 2022

Chapter 22

Orbital Dance

Through all the explorations of resources for people and planet, let us dance together to wake-up calls through each season of the year.

Orbital Dance

Just know,
Earth does its orbital dance by tilting on its axis,
Producing night and day in an elliptical or elongated circle fashion Creating Fall, Winter, Spring and Summer
With effortless ease as we play and as we slumber.

As she tilts around during the northern winter
The Sun's rays heat the South immensely
Creating the wonder of
Southern Summer, intensely.

And then Earth starts to orbits away from Summer
Now that her orbital dance is pointed completely
away from the sun,
Autumn has begun

The Earth keeps dancing by tilting around
During the northern summer
And the Sun's rays heat the North more intensely
Hence, the bitter of
Northern Winter.

Earth begins her tilts in an orbit Earth away from Winter
And before her orbital dance is pointed directly at the sun,
Spring is on.

As Earth orbits away from winter before it is pointed directly
at the sun, spring is created
As Earth orbits away from summer before it is pointed
completely away from the sun fall is created.

Whereas the Tropics face the Sun all year round
At the Equator,
The tilting Earth, a spinning creator.

Thus, it takes 365 days for the tilting dance to orbit around the
Sun
Creating Summer, Winter, Spring and Fall
Oh, what dancing fun!

©2015 by LaVerne McLeod

**Let's dance through each season being in action to help
make this a better world.**

The Crossroads of Social and Climate Justice

Bibliography

Chapter 1- Wake Up Calls

1. Douglass, Frederick. *Frederick Douglass: Portrait of a Free Man: The Narrative of the Life of Frederick Douglass,* with elucidations by Henry Louis Gates, Jr. New burgh, NY: Thornwillow Press, 2019.
2. "26 Thought Provoking Quotes by Chief Seattle." *The Famous People.* 2019 https://quotes.thefamouspeople. com/chief-seattle-1043.php>.
3. Killgrove, Kristina. "Medieval Cold Snap that Caused Famine and Death Reveals Danger of Climate Change." *Forbes.* December, 2016. https://www.forbes.com/ sites/kristinakillgrove/2016/12/01/medieval-cold-snap-that-caused-famine-and-death-reveals-danger-of-climate-change/?sh=541751905344.
4. Flanagan, Padraic. "The 15th Century Was So Cold People Used Fire to Melt Frozen Wine Bottles." *Associated Newspapers Ltd.* December 1, 2016. https:// inews.co.uk/news/science/coldest-decade-last-1000-years-offers-climate-change-lessons-34319
5. "The Doctrine of Discovery, 1493." *The Gilder Lehrman Institute of American History.* https://www.gilderleh rman.org/history-resources/spotlight-primary-source/ doc-trine-discovery-1493
6. Eanes de Zurara, Gomes. *The Chronicle of the Discorery and Conquest of Guinea. Cambridge University Press first published 1896; compilation version 2010.*
7. "Fossil Fuel." *Wikipedia.* August 2022, https://

en.wikipedia.org/wiki/Fossil_fuel#:~:text=The%20
theory%20that%20fossil%20fuels%20formed%20
from%20the,%22as%20early%20as%201757%20
and%20
certainly%20by%201763%22

8. Hsu, Chang S. and Robinson, Paul R. *Springer Handbook of Petroleum Technology* (2nd illustrated ed., p. 360). New York, NY: Springer Publishing Co., 2017

9. ClearIAS Team. "American Revolution-How America Overturned Monarchy and Became Independent, Rich, and Powerful." *ClearIAS World History Notes.* October 14, 2018. https://www.clearias.com/american-revolution/#:~:text=The%20significance%20of%20the%20American%20Revolution%201%20The,federal%20government%20and%20states.%20...%20More%20items...%20

10. Oliver, Mark. "10 Ways American Slavery Continued Long After the Civil War." *Listverse*, June 17, 2017. https://listverse.com/2017/06/21/10-ways-american-slavery-continued-long-after-the-civil-war/

11. Nittle, Nadra Kareem. "How the Black Codes Limited African American Progress After the Civil War." *History*. January 28, 2021. https://www.history.com/news/black-codes-reconstruction-slavery?li_source=LI&li_medium=m2m-rcw-history

12. Loew, James W. *Sundown Towns: A Hidden Dimension of American Racism.* New York, NY: The New Press, 2005.

13. Mott, Ashley. "Fact Check: Southern States Used Convict Leasing to Force Black People into Unpaid Labor." *USA Today.*, July 7, 2020. https://www.usatoday.com/story/news/factcheck/2020/07/07/fact-check-convict-leasing-forced-black-people-into-unpaid-labor/5368307002/

14. "Jim Crow Laws 1930s Introduction." *YouTube.* https://www.youtube.com/watch?v=_IfLzgaJOMU&t=2s

15. "Transcript of Congressional Testimony of Dr. James

Hansen, US Senate Committee on Energy and Natural Resources, 23 June 1988." *SeaLevel* and *SeanMunger.* June 23, 2016.   https://www.sealevel.info/1988_Hansen_Senate_Testimony.html & https://seanmunger.com/2016/06/23/one-speech-can-change-the-world-james-hansens-warning-on-climate-change/

16. Powell, John A. *Racing for Justice*. Bloomington, IN: Indiana University Press, 2015.

17. Biography.com Editors. "Emmett Till Biography." *Biography*. August 7, 2022. https://www.biography.com/crime-figure/emmett-till.

18. History Pod. "1st December 1955: Rosa Parks Refuses to Give Up her Seat on a Bus After the White Section Filled Up." *YouTube.* https://www.youtube.com/watch?v=fclfSac0bqQ

19. Anirudh. "10 Major Accomplishments of Martin Luther King Jr." *Learnodo-Newtonic*. July 5, 2014. https://learnodo-newtonic.com/martin-luther-king-jr-accomplishments.

20. Peters, Whit. "The Make Good Trouble Legacy of Representative John Lewis." *Yakima Herald-Republic* February 27, 2022,

21. "This Day in History: Four Black Schoolgirls Killed in Birmingham Church Bombing." *History.* September 15, 2021. https://www.history.com/this-day-in-history/four-black-schoolgirls-killed-in-birmingham

22. Siemaszko, Corky. "Birmingham Erupted into Chaos in 1963 as Battle for Civil Rights Exploded in South." *New York Daily News*, May 3, 2012,

23. "Assassination of Martin Luther King, Jr." *Stanford University the Martin Luther King, Jr. Research and Education Institute.* Undated. https://kinginstitute.stanford.edu/encyclopedia/assassination-martin-luther-king-jr

24. Haley, Alex. *Roots: The Saga of an American Family.* New York, NY: Doubleday, 1976.

25. Walker, Alice. "Alice Walker's Garden." *Alice Walker: The Official Website.* July 2022. https://alicewalkersgar

den.com.

26. Editors of Encyclopedia Britannica. "August Wilson American Dramatist. *Brittanica.* https://www.britannica.com/biography/August-Wilson.

27. Authors and Review Editors. "Special Report Global Warming Summary for Policymakers." *Intergovernmental Panel on Climate Change.* October 5, 2018. https://www.ipcc.ch/sr15/chapter/spm/

28. Somero, Dr. George of Hopkins Marine Station of Stanford University, interviewed by McLeod, LaVerne. February 10, 2014.

29. Authors and Review Editors. "Special Report Global Warming Summary for Policymakers." *Intergovernmental Panel on Climate Change.* October 5, 2018. https://www.ipcc.ch/sr15/chapter/spm/

30. "Nobel Peace Prize 2007 Intergovernmental Panel on Climate Change (IPCC) and Albert Arnold (Al) Gore." *United Nations.* https://www.un.org/en/about-us/nobel-peace-prize/ipcc-al-gore-2007

31. Stacy, Mary. "Review of the End of Nature by Bill McKibben." *My World.* March 22, 2002; https://kilchurn.blogspot.com/2002/03/review-of-end-of-nature-by-bill.html

32. McKibben, Bill. "About 350." *350.* https://350.org/about/

33. Thill, Mary. "Bill McKibben on Why 350 is the Most Important Number." *Adirondack Almanack.* Aug. 21, 2009. https://www.adirondackalmanack.com/2009/08/bill-mckibben-on-why-350-is-the-most-important-number.html

34. McKibben, Bill. "Global Warming's Terrifying New Math." *Rolling Stone.* July 19, 2012. https://www.rollingstone.com/politics/politics-news/global-warmings-terrifying-new-math-188550/

35. Thill, Mary. "Bill McKibben on Why 350 is the Most Important Number." *Adirondack Almanack.* Aug. 21, 2009. https://www.adirondackalmanack.com/2009/08/

bill-mckibben-on-why-350-is-the-most-important-num
ber.html

36. Winters, Justin. "One Earth, an Ambitious Plan to Slow Global Climate Change." Speech at the Bioneers Conference and the Leonardo DiCaprio Foundation Conference in San Rafael, CA. Octo. 22, 2018.

37. Greta Thunberg's speech to the United Nations https://www.usatoday.com/story/news/2019/09/23/ greta-thunberg-tells-un-summit-youth-not-forgive-cli mate-inaction/2421335001/

38. Weise, Elizabeth. "'How Dare You?' Read Greta Thunberg's emotional climate change speech to UN and world leaders." *USA Today.* Sept. 23, 2019.

39. "School Strike for Climate." *Wikipedia.* Aug. 8, 2022. https://en.wikipedia.org/wiki/School_strike_for_the climate

40. "File 20-27 September Climate strikes attendee numbers.svg." *Wikipedia.* https://en.wikipedia.org/wiki/ File:20–27_September_Climate_strikes_attendee_num bers.svg

41. Thunberg, Greta. "The disarming case to act right now on climate change." *TED.*: https://www.ted.com/talks/ greta_thunberg_the_disarming_case_to_act_right_now_ on_climate

42. Koley, Jonathan. "To prevent the next pandemic, it's the legal wildlife trade we should worry about." *National Geographic.* May 7. 2020 https://www.nationalgeographic.com/animals/2020/05/ to-prevent-next-pandemic-focus-on-legal-wildlife-trade/

43. "After Rare Sight in the Ganges, Dolphins Spotted in Bosphorus Strait Between Europe and Asia," *News 18.* April 27, 2020 https://www.news18.com/news/buzz/ after-rare-sight-in-the-ganges-dolphins-spotted-in-bos phorus-strait-between-europe-and-asia-2593903.html

44. "Airborne Nitrogen Dioxide Plummets Over China." *National Aeronautics and Space Administration.* Jan.-Feb. 2020. https://www.earthobservatory.nasa.gov/

images/146362/airborne-nitrogen-dioxide-plummets-over-china%20%20%20and%20%C2%A045

45. "Coronavirus lockdown leading to drop in pollution across Europe." *European Space Agency.* March 27, 2020. https://www.esa.int/Applications/Observing_the_Earth/Copernicus/Sentinel-5P/Coronavirus_lockdown_1 eading_to_drop_in_pollution_across_Europe

46. "Air Pollution Series of Scholarly Articles." *World Health Organization.* https://www.who.int/health-topics/air-pollution#tab=tab_1

47. Burke, Marshall. "COVID-19 reduces economic activity, which reduces pollution, which saves lives." *G-Feed: Global Food, Environment and Economic Dynamics.* March 8, 2020. http://www.g-feed.com/2020/03/covid-19-reduces-economic-activity.html?m=1

48. Koren, Marina. "The Pandemic is Turning the Natural World Upside Down," *The Atlantic.* April 27, 2020. https://www.theatlantic.com/science/archive/2020/04/coronavirus-pandemic-earth-pollution-noise/609316/

49. "Airborne Nitrogen Dioxide Plummets Over China." *National Aeronautics & Space Administration*, March 2, 2020. https://earthobservatory.nasa.gov/imges/146362/airborne-nitrogen-dioxide-plummets-over-china?utm=carousel

50. Fitz-Gibbon, Jorge. "Everything we know about the death of George Floyd." *New York Post.* May 20, 2020. https://nypost.com/2020/05/28/everything-we-know-about-the-death-of-george-floyd/

Chapter 2-The Risk Where We Live

1. Steinberg, Nik. "Assessing Exposure to Climate Risk in U.S. Municipalities." *Four Twenty-Seven.* May 22, 2018. https://427mt.com/2018/05/22/assessing-exposure-to-climate-change-in-us-munis/

2. "Sea Level Rise in New Jersey Projections and Impacts." *Rutgers NJ Climate Change Research Center.*

May 2020. https://njclimateresourcecenter.rutgers.edu/climate_change_101/sea-level-rise-in-new-jersey-projections-and-impacts/

3. Valentine, Katie. "The whitewashing of the environmental movement." *Grist.* Sept. 24, 2013. https://grist.org/climate-energy/the-whitewashing-of-the-environmental-movement/

4. "Cyclone vs. Hurricane." *Diffen.* https://www.diffen.com/difference/Cyclone_vs_Hurricane

5. "Hurricane Katrina." *History.* Aug. 9, 2019. https://www.history.com/topics/natural-disasters-and-environment/hurricane-katrina#failures-in-government-response

6. May, Sandra. "What Are Climate and Climate Change?" *National Aeronautics and Space Administration.* Aug. 7, 2017. https://www.nasa.gov/audience/forstudents/5-8/features/nasa-knows/what-is-climate-change-58.html

7. "Hurricane Harvey." *Wikipedia.* June 28, 2022. https://en.wikipedia.org/wiki/Hurricane_Harvey

8. Villalon, Jessica. "Flooding Disproportionately Affects People of Color." *Bayou City Waterkeeper.* Sept. 18, 2020. https://bayoucitywaterkeeper.org/flooding-disproportionately-impacts-people-of-color/

9. "Regional Health Impacts: Midwest." *Centers for Disease Control and Preventioin.* Feb. 25, 2021. https://www.cdc.gov/climateandhealth/effects/midwest.htm#:~:text=Increased%20daytime%20and%20nighttime%20temperatures%20are%20associated%20with,heat%20and%20access%20to%20care%20is%20a%20concern.

10. Gignac, James. "Extreme Heat is a Threat to Midwestern Outdoor Workers." *Union of Concerned Scientists.* Aug. 24, 2021. https://blog.ucsusa.org/james-gignac/extreme-heat-is-a-threat-to-midwestern-outdoor-workers/

11. "Wildfires." *Centers for Disease Control and Prevention.* June 18, 2020. https://www.cdc.gov/climateandhealth/effects/wildfires.htm.

12. Farbotko, Carol and Boas, Ingrid. "Where will you be

able to live in 20 years?" *YouTube*. https://www.youtube.com/watch?v=M3XZBYVSnJ0

13. Ayazi, Hossein, and Elsheikh, Elsadig. "Climate Refugees: The Climate Crisis and Rights Denied." *University of California at Berkeley, Othering & Belonging Institute*. Dec. 10, 2019. https://belonging.berkeley.edu/sites/default/files/climate_refugees.pdf?file=1&force=1

Chapter 3-Change the Narrative

1. Gildea, Kevin. "Impressive and challenging look at the history of racist ideas in the US>" *The Irish Times*. July 29, 2017. https://www.irishtimes.com/culture/books/impressive-and-challenging-look-at-the-history-of-racist-ideas-in-us-1.3165233

2. Coronado, Gary. "Native Americans seek to change the name Squaw Valley." *Fontana News Room*. https://fontananewsroom.com/native-americans-seek-to-change-the-name-squaw-valley/

3. Gehr, Emma. "Indian mascots are offensive to native culture and should be replaced." *University of Connecticut The Daily Campus*. Oct. 15, 2020. https://dailycampus.com/2020/10/15/indian-mascots-are-offensive-to-native-culture-and-should-be-replaced/

4. Sample, Ian. "He, she or --? Gender-neutral pronouns reduce biases-study." *The Guardian*. August 5, 2019. https://www.theguardian.com/science/2019/aug/05/he-she-or-gender-neutral-pronouns-reduce-biases-study

5. "Report Reveals Certain Amount of Global Warming Irreversible." *Earth Day*. Aug. 11, 2021. https://www.earthday.org/report-reveals-certain-amount-of-global-warming-irreversible/

6. "2020 Ties for Hottest Year on Record." *Earth Day*. Jan. 15, 2021. https://www.earthday.org/2020-ties-for-warmest-year-on-record/

7. "Water and health: safety, accessibility, sustainability." *The Lancet*. Sept. 3, 2020. https://www.thelancet.com/

journals/lancet/article/PIIS0140-6736(18)32594-7/fullt
ext

8. "Most of the Air Pollution We Breathe Indoors Comes
    from Outside." *Earth Day.* July 8, 2021. https://www.
    earthday.org/most-of-the-air-pollution-we-breathe-in
    doors-comes-from-outside/

9. "2021 Incident Archive." *CalFire.* https://www.fire.
    ca.gov/incidents/2021/

10. "The Calm Before the Storm? Experts Predict Hurricane
    Season to Get Worse." *Earth Day.* Oct. 8, 2019. https://
    www.earthday.org/the-calm-before-the-storm-experts-
    predict-hurricane-season-to-get-worse/

11. "North American Winters are Losing Snow and Ice."
    Earth Day. Oct. 10, 2019. https://www.earthday.org/
    north-american-winters-are-losing-snow-and-ice/

12. "U.N. Report: Nature's Dangerous Decline 'Unprec-
    edented'; Species Extinction Rates 'Accelerating'."
    United Nations Sustainable Development. May 6, 2019.
    https://www.un.org/sustainabledevelopment/
    blog/2019/05/nature-decline-unprecedented-report/

13. "Climate and Environmental Literacy. *Earth Day.*
    https://www.earthday.org/campaign/climate-environ
    mental-literacy/

Chapter 4- Measuring CO$_2$ and Racial Discrimination

1. "What is the Greenhouse Effect?" *National Aeronautics
    and Space Administration: Climate Kids.* https://
    climatekids.nasa.gov/greenhouse-effect/

2. "Overview of Greenhouse Gases." *U.S. Environmen-
    tal Protection Agency.* May 16, 2022. https://www.epa.
    gov/ghgemissions/overview-greenhouse-gases

3. "Keeling Curve." *Wikipedia.* Aug. 30, 2022. https://
    en.wikipedia.org/wiki/Keeling_Curve

4. "The Keeling Curve." *University of California at San
    Diego Scripps Institute of Oceanography.* https://
    scripps.ucsd.edu/programs/keelingcurve/

5. Scripps Oceanography. "How Scientists Measure Carbon Dioxide in the Air." *YouTube.* https://www.youtube.com/watch?v=dXBzFNEwoj8

6. "Can We Measure Racism? Yes We Can." *Replication Index.* June 11, 20202. https://replicationindex.com/2020/06/11/can-we-measure-racism-yes-we-can/

7. Omowote, Agbolade. "Research shows AI is often biased. Here's how to make algorithms work for all of us." *World Economic Forum.* July 19, 2021. https://www.weforum.org/agenda/2021/07/ai-machine-learning-bias-discrimination

8. Dastin, Jeffrey. "Amazon scraps secret AI recruiting tool that showed bias against women." *Reuters.* Oct. 10, 2018. https://www.reuters.com/article/us-amazon-com-jobs-automation-insight-idUSKCN1MK08G

9. Najibi, Alex. "Racial Discrimination in Face Recognition Technology." *Harvard University Graduate School of Arts and Sciences.* Oct. 24, 2020. https://sitn.hms.harvard.edu/flash/2020/racial-discrimination-in-face-recognition-technology/

10. Obermeyer, Ziad, *et al.* "Dissecting racial bias in an algorithm used to manage the health of populations." *Science.* Oct. 26, 2019. https://www.science.org/doi/10.1126/science.aax2342

11. Hawaiian Observatory Jobs. https://www.ziprecruiter.com/Jobs/Observatory/--in-Hawaii

12. Global Warming Mitigation Project. https:/www.globalwarmingmitigationproject.org/about-kcp

Chapter 5- Transportation

2. Anderson, Monica. "Who relies on public transit in the U.S." *Pew Research Center.* April 7, 2016. https://www.pewresearch.org/fact-tank/2016/04/07/who-relies-on-public-transit-in-the-u-s/

3. Kneebone, Elizabeth, and Holmes, Natalie. "The growing distance between people and jobs in metropoli-

tan America." *Metropolitan Policy Program at Brookings.* March 1, 2015. https://www.brookings.edu/wp-content/uploads/2016/07/Srvy_JobsProximity.pdf

4. Sanchez, Thomas W. "The Connection between Public Transit and Employment." *Journal of the American Planning Association.* Nov. 26, 2007. https://doi.org/10.1080/01944369908976058 The Connection Between Public Transit and Employment, by Thomas W. Sanchez.

5. English, Jonathan. "Why Public Transportation Works Better Outside the U.S." *Bloomberg.* Oct. 10, 2018. https://www.bloomberg.com/news/articles/2018-10-10/why-public-transportation-works-better-outside-the-u-s

6. "Transportation Sector Emissions." *Environmental Protection Agency.* Aug. 5, 2022/ https://www.epa.gov/ghgemissions/sources-greenhouse-gas-emissions#transportation

7. "Sustainability of Royal Caribbean Cruises." *The Costco Connection.* July 2016.

8. "Search Results—Electric Vehicles." *Environmental Protection Agency.* https://search.epa.gov/epasearch/?querytext=electric+vehicles&areaname=&areacontacts=&areasearchurl=&typeofsearch=epa&result_template=#/

9. Schmutzler, Aaron. "The hidden benefits of high-speed rail." *Nature.* Oct. 25, 2021. https://www.nature.com/articles/s41558-021-01199z#:~:text=However%2C%20the%20expansion%20of%20the%20high-speed%20rail%20system,the%20rail%20system%20is%20exclusively%20for%20passenger%20transportation.

10. "States Offering Driver's Licenses to Immigrants." *National Conference of State Legislatures.* Aug. 9, 2021. https://www.ncsl.org/research/immigration/states-offering-driver-s-licenses-to-immigrants.aspx

11. Thompson, Erin. "Fixing Transit Requires Grappling with Racism." *TransLoc.* July 9, 2021. https://transloc.com/blog/fixing-transit-requires-grappling-racism/

12. Hamilton, James. "Careers in Electric Vehicles." *U.S. Bureau of Labor Statistics.* https://www.bls.gov/green/electric_vehicles/#hybrid
13. "All short flights can be zero-emissions starting in 2026." *Wright Electric.* https://www.weflywright.com/
14. Pipistrel.https:www.piptrel-usa.com/

15. "EV100 Group." *Climate Group.* https://www.theclimategroup.org/ev100-members
16. "Green Vehicle Guide." *Environmental Protection Agency.* Aug. 1, 2022. https://www.epa.gov/greenvehicles
17. "Sources of Greenhouses Gases: Transportation Sector Emissions." *Environmental Protection Agency.* Aug. 5, 2022. https://www.epa.gov/ghgemissions/sources-greenhouse-gas-emissions#transportation

Chapter 6- Disposable World, Part I

1. Elliott, Jane. "Would you stand up?" https://www.rev.com/blog/transcripts/jane-elliott-speech-transcript-on-race-being-black-in-america
2. Hopkins, Hop. "Racism is Killing the Planet." *Sierra.* June 8, 2020, https://www.sierraclub.org/sierra/racism-killing-planet
3. "Air Pollution in the Western Pacific." *World Health Organization.* https://www.who.int/westernpacific/health-topics/air-pollution
4. Costley, Drew. "Coal Is in the Decline, but its Effects Still Ravage Black and Latinx Communities." *Future Human.* April 4, 2021. https://futurehuman.medium.com/coal-is-in-decline-but-its-effects-still-ravage-black-and-latinx-communities-3efd3a2bc6e0
5. "Most Recent Asthma State or Territory Data." *Centers for Disease Control and Prevention.* April 25, 2022. https://www.cdc.gov/asthma/most_recent_data_states.htm

6. "Quick Facts Richmond city, California." *United States Census Bureau.* https://www.census.gov/quickfacts/richmondcitycalifornia

7. Hammon, Alicia. "No Coal in Humboldt." *The North Coast Environmental Center.* Nov. 22, 2021. https://www.yournec.org/no-coal-in-humboldt/

8. Schuppe, Jon. "A power plant's expansion plan galls environmentalists—and sows dread in a Black enclave next door." *NBC News.* April 9, 2022. https://apple.news/ALlyQd83MS3WW3gwxE3nDPg

9. "Coal Blooded Action Toolkit." *NAACP.* https://naacp.org/resources/coal-blooded-action-toolkit

10. "Natural Gas Explained." *U.S. Energy Information Administration.* Dec. 8, 2021. https://www.eia.gov/energyexplained/natural-gas/natural-gas-and-the-environment.php

11. "Do closed coal mines emit methane?" *Environmental Protection Agency.* July 1, 2017. https://www.epa.gov/cmop/about-coal-mine-methane#q4

12. Shwe, Elizabeth. "Bipartisan Coal Transition Bill Withdrawn by House Sponsor." *Maryland Matters.* Mar. 8, 2021. https://www.marylandmatters.org/2021/03/08/bipartisan-coal-transition-bill-withdrawn-by-house-sponsor/

13. "CalEnviroScreen." June 25, 2018. *California Office of Environmental Health Hazard Assessment.* June 25, 2018. https://oehha.ca.gov/calenviroscreen/report/calenviroscreen-30

14. Arcury, Thomas and Quandt, Sara A. "Chronic Agricultural Chemical Exposure Among Migrant and Seasonal Workers." *National Library of Medicine.* Oct. 4, 2019. https://www.ncbi.nlm.nih.gov/pmc/articles/PMC6777717/

15. Gonzalez, Eduardo Jr. "Migrant Farm Workers: Our Nation's Invisible Population." *Diversity, Equity and Inclusion Community of Practice.* June 6, 2019. https://copdei.extension.org/migrant-farm-workers-our-nations-

invisible-population/ an article by Eduardo González, Jr. of Diversity, Equity & Inclusion Community of Practice

16. "Don't Forget the Elders during the Holidays." *Purple Feather Press.* https://conta.cc/328vzSb

17. Eastway, Monica. "ECO Generation Park the Vision." *YouTube.* Dec. 12, 2021. https://www.youtube.com/watch?v=YJ88OKaZ4v4   Eco Generation Park

18. Stotts, Isaak. "The War on Drugs and Racism: Its Effect on US Demographics." *Addiction Resource.* Jan. 6, 2022. https://addictionresource.com/guides/war-on-drugs/

19. History Editors. "Black Codes." *History.* Jan. 26, 2022. https://www.history.com/topics/black-history/black-codes#impact-of-the-black-codes

20. Camera, Lauren. "School Suspension Data Shows Glaring Disparities in Discipline by Race." *US News.* Oct. 13, 2020. https://www.usnews.com/news/education-news/articles/2020-10-13/school-suspension-data-shows-glaring-disparities-in-discipline-by-race

21. Georgia Appleseed Center for Law & Justice. "What is the School-to-Prison Pipeline? How Do We Dismantle It? *YouTube.* https://www.youtube.com/watch?v=XUUTDoWusHI&t=1s

22. Vice News. "New Ways Private Prisons Are Making Billions." *YouTube*: April 22, 2022. https://www.youtube.com/watch?v=U1U_xQVSpBE

23 Panona, Amber. "US Prison Population by Race." *World Atlas.* July 18, 2019. https://www.worldatlas.com/articles/incarceration-rates-by-race-ethnicity-and-gender-in-the-u-s.html

24. Harding, David, *et al.* "Incarceration." *Stanford Center on Poverty & Inequality.* https://inequality.stanford.edu/mobile/cpi-research/area/incarceration

25. Tiseo, Ian. "Plastic waste in the United States – statistics and faces." *Statistica.* Nov. 22, 2021. https://www.statista.com/topics/5127/plastic-waste-in-the-united-

states/

26. *Ibid.*
27. "Basic Information on Landfill Gas." *Environmental Protection Agency.* April 21, 2022. https://www.epa.gov/lmop/basic-information-about-landfill-gas
28. "Synthesis Report Climate Change." *Intergovernmental Panel on Climate Change (IPCC).* Oct. 2, 2014.
29. "Benefits and Cost for Migrating Methane Emissions." *UN Environment Programme Report.* May 2021. https://www.unep.org/resources/report/global-methane-assessment-benefits-and-costs-mitigating-methane-emissions
30. Katz, Cheryl. "The World's Recycling is in Chaos." *Wired.* Mar. 13, 2019. https://www.wired.com/story/the-worlds-recycling-is-in-chaos-heres-what-has-to-happen/
31. "The Coalition of Homeless Service Providers." Monterey County. 2022 Directory https://homeless.chsp.org/
32. "Benefits of Restorative Justice." *Community Justice Network of Vermont.* http://cjnvt.org/about-us/benefits-of-restorative-justice/
33. Hampshire PCC. "What is Restorative Justice?" *YouTube.* Nov. 19, 2016. https://www.youtube.com/watch?v=ZfQhfN6PxPI
34. Thompson, Andrew. "Understanding the Meaning of Ubuntu: A Proudly South African Philosophy." *Culture Trip.* June 11, 20202. https://theculturetrip.com/africa/south-africa/articles/understanding-the-meaning-of-ubuntu-a-proudly-south-african-philosophy/
35. "In the Field: Where We Work." *Restorative Justice Exchange.* https://restorativejustice.org/where/
36. "Racial Healing Tool Kit Activities for Youth Involvement." *W.K. Kellogg Foundation.* https://healourcommunities.org/wp-content/uploads/2022/06/Students-Action-Kit_2022-2.pdf
37. Georgia Appleseed Center for Law & Justice. "What is the School-to-Prison Pipeline? How Do We Dismantle It? *YouTube.* https://www.youtube.com/watch?v=XUUTDoWusHI&t=1s

38. Bailey, Richard, *et al.* "Breaking the Plastic Wave." *Pew Charitable Trusts.* 2020. https://www.pewtrusts.org/-/media/assets/2020/07/breakingtheplasticwave_summary.pdf

39. "Preventing Ocean Plastics." *The Pew Charitable Trusts.* July 2020. www.pewtrusts.org/en/projects/Preventing-Ocean-Plastics.

40. "Ocean Plastics." SystemIQ. Oceanplastics@systemiq.earth 2022.

41. "SB 1383 Education and Outreach Resources." CA.gov. January 1, 2022. https://calrecycle.ca.gov/organics/slcp/education/

42. "California's Short-Lived Climate Pollutant Reduction Strategy." *California Department of Resources Recycling and Recovery (CalRecycle).* www.calrecycle.ca.gov/organics/slcp

Chapter 7—Disposable World, Part II – The Enneagram Solution Story, Unlocking Your Inner Prison

1. Riso, Don Richard and Hudson, Russ. *The Wisdom of the Enneagram: The Complete Guide to Psychological and Physical Growth for the Nine Personality Types.* New York, NY: Bantam Books, 1999.

2. "What We Do." *Enneagram Prison Project.* https://enneagramprisonproject.org/what-we-do/

3. "9 PrisonsOneKey," a virtual enneagram course. Enneagram Prison Project, EPP. https://enneagramprisonproject.org/9prisonsonekey-training/

Chapter 8 -Deforestation

1. Leahy, Stephen. "Tropical Forest Loss Slowed in 2017—To the Second Worst Total Ever." *National Geographic.* June 27, 2018. https://www.nationalgeographic.com/news/2018/06/tropical-deforestation-forest-loss-2017/

2. "The State of the World's Forests 2020." https://www.fao.org/state-of-forests/en/
3. Fearnside, Philip. "Business as Usual: A Resurgence of Deforestation in the Brazilian Amazon." *Yale Environment 360.* April 18, 2017. https://e360.yale.edu/features/business-as-usual-a-resurgence-of-deforestation-in-the-brazilian-amazon
4. "Global Forest Atlas." *Yale School of the Environment.* 2020. "Forest Regions." The Global World Atlas, Yale University 2020. https://yff.yale.edu/current-projects/global-forest-atlas
5. Gordon, Graham. "Addressing Deforestation in Indonesia." *The Borgen Project.* Nov. 1, 2019. https://borgenproject.org/tag/deforestation-in-indonesia/
6. Mega, Emiliano Rodriguez. "Wild coffee species threatened by climate change and deforestation." *Nature.* Jan. 16, 2019. https://www.nature.com/articles/d41586-019-00150-9
7. "Climate Change." *Stop Clearcutting California.* https://www.stopclearcuttingca.org/contact/
8. "Fighting Deforestation and Illegal Logging." *Environmental Investigation Agency.* https://eia-international.org/forests/fighting-deforestation/
9. "Trees for the Future." *Facebook.* https://www.facebook.com/TreesfortheFuture/?ref=ts&fref=ts
10. *Ibid.*
11. Saunders, Steve. Personal interview. Big Sur, CA. 2019
12. "Our Impacts." *Rainforest Alliance.* https://www.rainforest-alliance.org/impact
13. "Join Our Team." *Rainforest Alliance.* https://www.rainforest-alliance.org/careers
14. "Stabilizing Our Climate by Protecting and Restoring Nature." *Conservation International.* https://www.conservation.org/priorities/forests
15. "Conservation." *Taleo.* https://chu.tbe.taleo.net/chu04/ats/careers/v2/searchResults?org=CONSERVATION&cws=39

16. "Choose Forests." *Forests Forever.* https://us.fsc.org/en-us

17. "Find the Frog." *Rainforest Alliance.* https://www.rainforest-alliance.org/find-certified/

18. Pachamama Alliance, https://pachamama.org/about/contact

Chapter 9- Air as Renewable Energy

1. "Renewable Energy." Office of Energy Efficiency and Renewable Energy. https://www.energy.gov/eere/renewable-energy

2. "What is Wind Power" *Acciona.* https://www.acciona.com/renewable-energy/wind-power/

3. Andersen, Allan D.; Hanson, Jens; Normann, Hakon E.; and Thune, Taran M. "The Norwegian oil and gas sector's influence on offshore wind power." *Journal of Cleaner Production.* Vol. 177, Centre for Technology, Innovation and Culture, University of Oslo. 2018. 813-823

4. Mäkitie, Ruukka; Normann, Hakon E.; Thune, Taran M; and Gonzales, Jakoba Sraml. "The green flings: market fluctuations and incumbent energy industries' engagement in renewable energy". *Centre for Technology, Innovation and Culture, University of Oslo.* 2018. https://ideas.repec.org/s/tik/inowpp.html

5. "Utilizing Expertise from Offshore Oil and Gas for Offshore Wind." *Business Norway.* Nov. 22, 2019.

6. "Equinor Wind Farm." *Equinor.* https://www.equinor.com/en/media-centre.html#media-contacts

7. "Alta Wind Energy Center." *Terra-Gen Power.* July 25, 2014. https://www.power-technology.com/projects/alta-wind-energy-center-awec-california/

8. "Los Vientos Windpower." *Duke Energy.* https://www.duke-energy.com/Our%20Company/About%20Us/Businesses/Renewable%20Energy/Wind%20Energy/Shirley%209.Windpower%20Contact%20Us

"Wind Turbine Manufacturers." *Energy Acuity.*
https://energyacuity.com/blog/2019-top-10-wind-turbine-manufacturers/)

Chapter 10

1. "Atmospheric River Portal." *National Oceanographic and Atmospheric Administration Physical Sciences Laboratory.* http://www.esrl.noaa.gov/psd/atmrivers/
2. Benyus, Janine. "Biomimicry." Lecture at Bioneers Conference Center, San Rafael, CA. 2016
3. "Humanity's Biggest Challenges. Nature's Human Solutions." *Biomimicry Institute.* https://biomimicry.org
4. "Oceanic Institute." *Hawaii Pacific University.* https://www.hpu.edu/oi/
5. Somero, Dr. George, Professor of Marine Science Emeritus (Stanford University) personal interview at Hopkins Marine Station in Pacific Grove, CA. 2014.
6. Berwyn, Bob. "Global Warming Is Messing with the Jet Stream. That Means More Extreme Weather." *Inside Climate News.* Oct. 31, 2018. https://insideclimatenews.org/news/31102018/jet-stream-climate-change-study-extreme-weather-arctic-amplification-temperature?gclid=EAIaIQobChMI6dm_heWH6wIVDh6tBh1LsgHXEAAYASAAEgKPOvD_BwE
7. "Background on: Hurricane and windstorm deductibles." *Insurance Information Institute.* https://www.iii.org/article/background-on-hurricane-and-windstorm-deductibles
8. "GW Researchers: 2,975 Excess Deaths Linked to Hurricane Maria. *GW Today-George Washington University.* Aug. 29, 2018. https://gwtoday.gwu.edu/gw-researchers-2975-excess-deaths-linked-hurricane-maria.
9. Bhatia, Kieran. "Global Warming and Hurricanes." *National Oceanographic and Atmospheric Administrtion, Geophysical Fluid Dynamics Laboratory.* July 12, 2022. https://www.gfdl.noaa.gov/global-warming-and-

hurricanes/

10. *Ibid.*

11. Dow, Kirstin, and Downing, Thomas E. *The Atlas of Climate Change.* Berkeley, CA: University of California Press, 2006.

12. *Ibid.*

13. MacCowan, Richard. "Biomimicry Beyond Organisms: Acting and Informing at a Systems-Level." PhD project scholarly document.

14. *Ibid.*

15. Farnsworth, Margo. "Stopping Floods with Nature's Help." *Tree Hugger.* June 12, 2013. treehuger.com June 12, 2013.

16. Henson, Paul and Usher, Donald J. *The Natural History of Big Sur*, Berkeley, CA: University of California Press, 1993. 134-135

17. Ingels, Bjarke. "Floating cities, the LEGO House and other architectural forms of the future." *TED.* 2019. https://www.ted.com/talks/bjarke_igels_floating_cities_the_lego_house_and_other_archi tectural_forms_of_the_future?utm_source=newsletter_daily&utm_campaign=daily&utm_medium=email&utm_content=button__2019-06-03

18. Further research Biomimicry https://biomimicry.org/what-is-biomimicry/ or https://asknature.org/

19. "Opportunities: Become Involved with MAP. *Mangrove Action Project.* https://mangroveactionproject.org/oppor tunities/

20. "Contact." *Global Mangrove Alliance.* http://www.man grovealliance.org/contact/

21. "Committed to Sustainability and Climate-Positive Actions." *Tetra Tech.* https://www.tetratech.com

Chapter 11- Coral Reef Disruptions

1. "Coral reef." *Wikipedia.* Aug. 29, 2022. https://en.wikipedia.org/wiki/Coral_reef

2. Jamie. "Different Types of Coral Reefs." *Different Types.* Dec. 17, 2020. https://www.differenttypes.net/types-of-coral-reefs/

3. "What is Ocean Acidification?" *National Oceanographic and Atmospheric Administration PMEL Carbon Pro gram.* https://www.pmel.noaa.gov/co2/story/

4. Albright, Rebecca; Mason, Benjamin; Miller, Margaret and Langdon, Chris. "Ocean acidification compromises recruitment success of the threatened Caribbean coral *Acropora palmate.*" *Proceeding of the National Academy of Sciences.* November 23, 2010 https://doi.org/10.1073/pnas.1007273107.

5. "Coral Reefs." *Environmental Protection Agency.* June 21, 29022, https://www.epa.gov/coral-reefs/

6. "Coral reefs: Essential and Threatened." *National Oceanographic and Atmospheric Administration.* April 14, 2016. https://www.noaa.gov/explainers/coral-reefs-essential-and-threatened

7. "Great Barrier Reef." *National Geographic.* https://education.nationalgeographic.org/resource/great-barrier-reef

8. "Human impact of coral bleaching in the Great Barrier Reef." *UK Essays.* Nov. 1, 2018. https://www.ukessays.com/essays/geography/human-impacts-of-coral-bleaching-in-the-great-barrier-reef.php

9. Papenfuss, Mary. "Swim Caps for Black Hair Banned at Olympics Because They Don't Fit 'Natural' Head Shape." *Huffington Post.* July 5, 20201. https://www.huffpost.com/entry/soul-cap-olympics-fina-ban_n_60e22c34e4b08f6f784bf0e7

10. *Ibid.*

11. "International Swimming Federation Reconsiders Afro Swim Cap After Backlash." The Black Wall Street Journal-Sports. Erika DuBose. July 7, 2021.

12. "Welcome to WIOMSA." *Western Indian Ocean Marine Science Association..* https://www.wiomsa.org/

13. "News and blogs." *Nature Seychelles.* http://natureseychelles.org/ & http://natureseychelles.org/get-involved/

jobs

14. "Giving coral reefs a future." *Secore International* http://www.secore.org/site/home.html & http://www. secore.org/site/our-work.html & careers: http://www. secore.org/site/about-us.html

15. `"Projects List." *Pur Projet.* https://www.purprojet.com/ project-list/ & https://www.purprojet.com/recruitment/

16. "The Reef." *Great Barrier Reef Foundation.* https:// www.barrierreef.org/

Chapter 12- Ice Sheets Melting

1. University of Cambridge. "Accelerating melt rate makes Greenland Ice Sheet world's largest 'dam'." *Science Daily.* Feb. 21, 2022. https://www.sciencedaily. com/releases/2022/02/220221155200.htm

2. Dow, Kristin, and Downing, Thomas. *The Atlas of Climate Change.* Berkeley, CA: University of California Press 2006 (revised).

3. Byrne, Kevin. "Trillion-ton iceberg breaks off from Antarctic ice shelf." *AccuWeather.* July 12, 2017. https:// www.accuweather.com/en/weather-news/trillion-ton-iceberg-breaks-off-from-antarctic-ice-shelf/358864

4. "NASA Study." *Earth and Planetary Science Letters.* May 14, 2015. https://www.sciencedirect.com/journal/ earth-and-planetary-science-letters

5. Climate Central Map. https://coastal.climatecentral.org/

6. Angakkorsuaq, Angaangaq. "Conditions of the Environment." *Ice Wisdom.* https://icewisdom.com/ blog/2020/12/angaangaq-about-the-conditions-of-the-environment/

7. "Permafrost." *Wikipedia.* Aug. 25, 2022. https:// en.wikipedia.org/wiki/Permafrost

8. "Glacier and Permafrost Hazards." *Arctic Program.* Jan. 5, 2022. https://arctic.noaa.gov/Report-Card/Report-Card-2021/ArtMID/8022/ArticleID/951/Glacier-and-Permafrost-Hazards

9. "Research: Polar Bears and the Changing Arctic." *Polar Bear International.* https://polarbearsinterna tional.org/research/see-our-projects/
10. "The Urgency in Alaska." *Alaska Conservation Foundation https://alaskaconservation.org/.*
11. "What We Fund: Places." *Climate Justice Resilience Fund.* https://www.cjrfund.org/places
12. Angakkorsuaq, Angaangaq. "Conditions of the Environment." *Ice Wisdom.* https://icewisdom.com/ blog/2020/12/angaangaq-about-the-conditions-of-the-environment/
13. "Let's Create a Better World: Bridge Building to Equity Workshops." *LaVerne McLeod.* https://www.lavernem cleod.com

Chapter 13-Water as Renewable Energy

1. "How Hydropower Works." *Office of Energy Efficiency and Renewable Energy.* https://www.energy.gov/eere/ water/how-hydropower-works
2. "Facts About Hydropower." *Wisconsin Valley Improvement Company.* http://www.wvic.com/content/facts_ about_hydropower.cfm
3. Bhattacharya, Anirudra. "Top Ten Hydropower Companies." *LinkedIn.* May 1, 2016. https://www.linkedin. com/pulse/top-10-hydropower-companies-anirudra-bhat tacharya/
4. "Hydropower Companies (Hydro Energy)." *Xprt Energy.* https://www.energy-xprt.com/hydro-energy/hydro power/companies

Chapter 14- Heat Waves and Drought

1. "Desertification—A Brief Overview." *Byjus Exam Prep.* https://byjus.com/free-ias-prep/desertification/
2. University of Toronto. "Global warming is changing

organic matter in soil." *Phys.org.* Nov. 24, 2019. https://phys.org/news/2008-11-global-soil.html#jCp

3.   Dahl, Kristina, *et al.* "Increased frequency of and population exposure to extreme heat index days in the United States in the 21st century." *Environmental Research Communications.* July 16, 2019.

4.   Senesac, Emily. "The Chicago 1995 Heat Wave." *National Oceanographic and Atmospheric Administration.* https://vlab.noaa.gov/web/nws-heritage/-/great-chicago-heat-wave-of-1995

5.   Sidik, Saima May. "Estimating Heat Wave Frequency and Strength: A Chicago Case Study." *Eos: Science News by AGU.* March 10, 2022. https://eos.org/research-spotlights/estimating-heat-wave-frequency-and-strength-a-chicago-case-study

6.   Di Liberto, Tom. "Record-breaking June 2021 heatwave impacts the U.S. West." *National Oceanographic and Atmospheric Administration.* June 23, 2021. https://www.climate.gov/news-features/event-tracker/record-breaking-june-2021-heatwave-impacts-us-west

7.   Gatopoulos, Derek. "Southeast Europe heat wave set to be among the worst in decades." *Associated Press.* July 29, 2021 https://www.wane.com/news/southeast-europe-heat-wave-seen-as-among-worst-in-decades/

8.   "Climate Change Indicators: Heat Waves." *Environmental Protection Agency.* July 1, 2022. https://www.epa.gov/climate-indicators/climate-change-indicators-heat-waves

9.   "Chapter 6: Temperature Changes in the United States." *U.S. Global Change Research Program.* https://science2017.globalchange.gov/chapter/6/

10. "Climate Change Indicators: Heat-Related Deaths." *Environmental Protection Agency.* https://www.epa.gov/climate-indicators/climate-change-indicators-heat-related-deaths#ref21

11. Yu, Zhaowu, *et al.* "The cooling effect of urban green vegetation in different climate zones." *Research Gate.* April 30, 2019. https://www.researchgate.net/project/

The-cooling-effect-of-urban-green-vegetation-in-differ
ent-climate-zones

12. "Chicago, IL, Adapts to Improve Extreme Heat Preparedness." *Environmental Protection Agency.* Mar.
14, 2022. https://www.epa.gov/arc-x/chicago-il-adapts-
improve-extreme-heat-preparedness

13. Hayhoe, Katharine, *et al.* "Climate Change, Heat
Waves, and Mortality Projections for Chicago."
Journal of Great Lakes Research. July 1, 2010. https://
bioone.org/journals/journal-of-great-lakes-research/
volume-36/issue-sp2/j.jglr.2009.12.009/Climate-
Change-Heat-Waves-and-Mortality-Projections-for-
Chicago/10.1016/j.jglr.2009.12.009.short

14. "Excessive Heat Events Guidebook in Brief." *Environmental Protection Agency.* June 1, 2006. https://www.
epa.gov/sites/default/files/2014-07/documents/eheguide-
brief_final.pdf

15. Center for Climate and Energy Solutions. *https://www.
c2es.org/content/heat-waves-and-climate-change/*

16. Ventimiglia, Andrea. "Heatwaves." *Future Earth.* https://
futureearth.org/publications/issue-briefs/heatwaves/

17. "Work Heat Action Plan." *Heat Shield.* https://www.
heat-shield.eu/

18. "How to Prepare for a Drought." *WikiHow.* Jan. 21,
2022. https://www.wikihow.com/Prepare-for-a-Drought

Chapter 15- Food and Farming

1. Reich, Robert. "How America Created Its Shameful
Wealth Gap." *YouTube.* https://www.youtube.com/
watch?v=9diZJks95Ko

2. "The Black Codes and Jim Crow Laws." *World Atlas.*
https://www.worldatlas.com/articles/the-black-codes-
and-jim-crow-laws.html

3. "Welcome to Wikipedia." *Wikipedia.* en.wikipedia.org

4. "The Doctrine of Discovery, 1493." *Gilder Lehrman
Institute of American History.* https://www.gilderlehr

man.org/history-resources/spotlight-primary-source/
doctrine-discovery-1493

5. Guzman, Andrea and McDaniel, Piper. "This Land is Not Your Land." *Mother Jones.* May-June, 2021.

6. Taylor, Elizabeth Berlin. "Perspectives on the Trail of Tears." *Gilder Lehrman Institute of American History* https://www.gilderlehrman.org/history-resources/lesson-plan/perspectives-trail-tears

7. "The Valley Spotlight." https://rgvaff.com/series_history.html

8. Levin, Sam. "'This is all stolen land': Native Americans want more than California's Apology." *The Guardian.* June 21, 2019. https://www.theguardian.com/us-news/2019/jun/20/california-native-americans-governor-apology-reparations

9. Guzman, Andrea and McDaniel, Piper. "This Land is Not Your Land." *Mother Jones.* May-June, 2021.

10. Hannah-Jones, Nikole. *The 1619 Project: A New American Origin Story.* New York, NY. New York Times Magazine. 2021.

11. Myers. Barton. "Sherman's Field Order No. 15." *New Georgia Encyclopedia.* Sept. 30, 2020. https://www.georgiaencyclopedia.org/articles/history-archaeology/shermans-field-order-no-15/

12. "The Assassination of Abraham Lincoln." *U.S. History Online Textbook.* https://www.ushistory.org/us/34f.asp

13. "Records of Rights: Rights of Native Americans." *National Archives.* http://recordsofrights.org/records/285/indian-appropriations-act

14. White, Monica. *Freedom Farmers*: *Agricultural Resistance and the Black Freedom Movement, 1880-2010.* Chapel Hill, NC: The University of North Carolina Press, 2019.

15. "Alien Land Laws in California (1913 & 1920). *Immigration and Ethnic History Society.* https://immigrationhistory.org/item/alien-land-laws-in-california-1913-

16. "The Conservation Legacy of Theodore Roosevelt." *US

Department of the Interior Feb. 14, 2020. https://www. doi.gov/blog/conservation-legacy-theodore-roosevelt

17. "Alien Land Laws in California (1913 & 1920). *Immigration and Ethnic History Society.* https://immigrationhistory.org/item/alien-land-laws-in-california-1913-

18. Guzman, Andrea and McDaniel, Piper. "This Land is Not Your Land." *Mother Jones.* May-June, 2021.

19. History.com Editors. "New Deal." *History.* Oct. 5, 2021. https://www.history.com/topics/great-depression/new-deal

20. "Attack on Pearl Harbor." *The Center for Legislative Archives.* July 25, 2019. https://www.archives.gov/legislative/features/pearl-harbor

21. Daniel, Peter. *Dispossession, Discrimination against African American Farmers in the Age of Civil Rights.* Chapel Hill, NC: University of North Carolina Press, 2015.

22. U.S. District Court for the District of Columbia, Pigford versus Glickman and the U.S. Department of Agriculture, Jan. 5, 1999. https://www.dm.usda.gov/pigford.pdf

23. Rosenberg, Nathan, and Stucki, Bryce Wilson. "How the USDA distorted data to conceal decades of discrimination against Black farmers." *The Counter.* June 26, 2019. https://thecounter.org/usda-black-farmers-discrimination-tom-vilsack-reparations-civil-rights/

24. Guzman, Andrea and McDaniel, Piper. "This Land is Not Your Land." *Mother Jones.* May-June, 2021.

25. *Ibid.*

26. De Marinis, Jacopo. "What is the Justice for Black Farmers Act and What Will It Do?" *Chicago Area Peace Action.* Jan. 14, 2022. http://www.chipeaceaction.org/2022/01/14/what-is-the-justice-for-black-farmers-act-and-what-will-it-do/

27. "S.300: Justice for Black Farmers Act of 2021." *Gov Track.* https://www.govtrack.us/congress/bills/117/s300

28. Duff-Brown, Beth. "In California, the pandemic hits Latinos hard." *Stanford Medicine News Center.* May 13,

2021. https://med.stanford.edu/news/all-news/2021/05/in-california-the-pandemic-hits-latinos-hard.html

29. Guidi, Ruxandra. "Farmworkers face illness and death in the fields." *High Country News.* Aug. 20, 2018. https://www.hcn.org/issues/50.14/agriculture-californias-farm workers-face-illness-and-death-in-the-fields

30. Al-Kaisi, Mandi, *et al.* "Frequent tillage and its impact on soil quality." *Iowa State University Extension and Outreach.* June 28, 2004. https://crops.extension.iastate.edu/encyclopedia/frequent-tillage-and-its-impact-soil-quality

31. University of Toronto. "Global Warming Is Changing Organic Matter In Soil: Atmosphere Could Change As A Result." ScienceDaily. ScienceDaily, 28 November 2008. ww.sciencedaily.com/releas es/2008/11/081124130948.ht g

32. Ewing, Bob. "Global Warming Changing Organic Matter in Soil," *Nature Geoscience Journal*, Nov. 24, 2008.

33. Berardelli, Jeff. "'The world from our childhood is no longer here': Report details drastic changes as Arctic warms." *CBS News.* Dec. 10, 2019. https://www.cb snews.com/news/climate-change-noaa-arctic-re port-2019/

34. Turetsky, M.R , *et al.* "Carbon release through abrupt permafrost thaw." Published: 03 February 2020 in *Nature Geoscience*, Feb. 3, 2020.

35. "Recent ERS relating to rice." *United States Department of Agriculture (USDA).* Oct. 19, 2021. https://www.ers.usda.gov/topics/crops/rice.aspx#otherpublications

36. Bingner, Ronald and Locke, Martin. "Water Quality and Ecology Research: Oxford, MS." *United States Department of Agriculture (USDA),* Sept. 2, 2022. https://www.ars.usda.gov/research/project/?accnNo=433580

37. "Sustainable Solutions." *International Rice Research I nstitute.* https://www.irri.org/our-work/outcome-themes/developing-environmentally-sustainable-solutions-rice-systems

38. "Are Cows the Cause of Global Warming?" *Time for Change.* https://timeforchange.org/are-cows-cause-of-global-warming-meat-methane-CO2/

39. Applegate, Todd, *et al.* "Ag 101." *Environmental Protection Agency and Purdue University.* July 29, 2015. https://www.epa.gov/sites/production/files/2015-07/documents/ag_101_agriculture_us_epa_0.pdf

40. Water Resources. "Livestock water use." *United States Geological Survey, ((USGA).* Mar. 1, 2019. https://www.usgs.gov/mission-areas/water-resources/science/livestock-water-use?qt-science_center_objects=0#qt-science_center_objects

41. Hallock, Betty. "To Make a Burger You Need 660 Gallons of Water." *Los Angeles Times.* Jan. 27, 2014. https://www.latimes.com/food/dailydish/la-dd-gallons-of-water-to-make-a-burger-20140124-story.html

42. Spokes, Jessica. "The Effects of Factory Farming." *Animal Rights,* Mar. 21, 2011. https://sumsc.wordpress.com/2011/03/21/against-factory-farming/

43. "We Create Successful Organic Farms and Gardens Within Businesses, Schools and Institutions." *Grow Your Lunch.* https://www.growyourlunch.com/

44. Bernardi, Kari. "Making Healthy Living Fresh, Simple and Delicious." *Supernatural Chef.* http://www.supernaturalchef.com/

45. Rogers, Raphael E. "Covering Up Slavery in the Birth of the U.S." *National African-American Reparations Commission.* https://reparationscomm.org/

46. Levin, Sam. "'This is all stolen land.' Native Americans want more than California's apology." *The Guardian.* June 21, 2019. https://www.theguardian.com/us-news/2019/jun/20/california-native-americans-governor-apology-reparations

47. Reiley, Laura and Mara, Melina. "How California's Salinas Valley went from covid hot spot to a model for vaccination and safety." *The Washington Post.* June 6, 2021. https://www.washingtonpost.com/business/inter

active/2021/salinas-farmworker-covid-vaccinations/

48. Kutchta, David. "What is regenerative agriculture?" *Treehugger.* Feb. 17, 2021. https://www.treehugger.com/what-is-regenerative-agriculture-511285

49. Cummins, Ronnie. "Regeneration 2019: State of the Movement." *Regenerative Agriculture.* May 22, 2019. https://www.organicconsumers.org/blog/regeneration-2019-state-movement

50. Popkin, Gabriel and Demczuk, Gabriella. "Planting Crops & Carbon Too." *The Washington Post.* Jan. 22, 2021

51. "Case Study: Silvopastoralism." *Compassion in World Farming.* https://www.ciwf.org.uk/media/7430275/case-study-6-silvopastoral-systemspdf_87238.pdf

52 . Monterey County Weekly, "The Future of Farming? By Larissa Zimberoff, p.18, October 14-20, 2021

53. "Juicy Tomatoes in Winter? Thank Hydroponics." Ohio State University, Food and Agriculture. February 2020. https://insights.osu.edu/food/what-hydroponics

Chapter 16. Part II Food and Farming-Aquaponics

1. Wright, Justin; Galvin, James; and Ratzlaff, Janna, board of directors of the Both Company. Personal interview in author's home studio in Big Sur, CA, Nov. 9, 2015. CA.

2. "Aquaponics." *Wikipedia.* https://en.wikipedia.org/wiki/Aquaponics

3. "Make Business a Force for Good." *B-Labab Global.* www.bcorporation.net

4. "Aquaponics." *Indeed.* https://www.indeed.com/q-Aquaponic-jobs.html

5. "Career Opportunities in Aquaponics." *Nelson and Pade.* 2018. https://aquaponics.com/wp-content/uploads/2018/08/Careen-Opportunities-in-Aquaponics.pdf

6. "Get to Know Aquaponics." *Backyard Aquaponics.* http://www.backyardaquaponics.com/

7. "What is Aquaponics?" *The Aquaponic Source.* http://theaquaponicsource.com/what-is-aquaponics/
8. "For All Your Commercial Aquaponic Needs." *Aquaponic Solutions.* http://www.aquaponic.com.au/
9. Bernstein, Sylvia. *Aquaponic Gardening: A Step-By-Step Guide to Raising Vegetables and Fish Together.* Gabriola Island, BC, Canada: New Society Publishers, 2011.

Chapter 17-Biomass- A Renewable Energy

1. "Biomass." *Wikipedia.* https://en.wikipedia.org/wiki/Biomass
2. U.S. Department of Energy. "Biomass for Electricity Generation." *Whole Building Design Guide.* Sept. 15, 2015. https://www.wbdg.org/resources/biomass-electricity-generation
3. "Biomass Energy Basics." *National Renewable Energy Laboratory.* https://www.nrel.gov/research/re-biomass.html
4. "For lovers of meditation." *The Guided Meditation Site.* https://www.the-guided-meditation-site.com/
5. U.S. Bureau of Labor Statistics-Careers in Biofuels. https://www.bls.gov/green/biofuels/biofuels.htm 5
6. Office of Energy Efficiency & Renewable Energy-STEM and Education. https://www.energy.gov/eere/education/eere-stem-and-education
7. GE Renewable Energy Careers for students. https://jobs.gecareers.com/renewableenergy/global/en/students-graduates
8. GE Renewable Energy Careers for professionals. https://jobs.gecareers.com/renewableenergy/global/en/professionals

Chapter 18-Fires

1. "The climate loop: 6 ways global warming is fueling US

fires." *World Economic Forum.* Sept. 24, 2020. https://www.weforum.org/agenda/2020/09/climate-feedback-climate-change-forest-fires/

2. "Soberanes Fire Incident." *CalFire.* Oct. 13, 2016. https://www.fire.ca.gov/incidents/2016/7/22/soberanes-fire/

3. Burns, Dennis, analyst for the Soberanes fire, Big Sur, CA, August 2016.

4. Li, David K., and Johnson, Alex. "20,00-acre wildfire all but destroys Paradise, California." *NBC News.* Nov. 8, 2018. https://www.nbcnews.com/storyline/western-wildfires/wildfire-chars-more-5-000-acres-rural-northern-california-n934086

5. Redd, Candace. How wildfires disproportionately affect people of color." *University of Washington.* Sept. 2, 2021. https://urban.uw.edu/news/how-wildfires-disproportionately-affect-people-of-color/

6. "Plant trees and save lives." *Trees for the Future.* https://trees.org/

7. "Trees for the Future." *Facebook.* https://www.facebook.com/TreesfortheFuture/?ref=ts&fref=ts

8. "The Grove: Where We Plant." *One Tree Planted.* https://onetreeplanted.org/

9. "Indigenous Peoples Burning Network." *The Nature Conservancy.* http://www.conservationgateway.org/ConservationPractices/FireLandscapes/Pages/IPBN.aspx

10. "Fighting Fire with Fire." *The Nature Conservancy.* https://www.natureaustralia.org.au/what-we-do/our-priorities/climate-change/climate-change-stories/fighting-fire-with-fire/

11. "Job Opportunities." *Conservation Biology Institute.* https://consbio.org/general/pages/job-opportunities

12. Hessburg, Paul. "Why wildfires have gotten worse – and what we can do about it." *TED.* https://www.ted.com/talks/paul_hessburg_why_wildfires_have_gotten_worse_and_what_we_can_do_about_it

Chapter 19- Solar as Renewable Energy

1. "Planning a Home Solar Electric System." *U.S. Department of Energy.* https://www.energy.gov/energysaver/planning-home-solar-electric-system
2. Huxley-Reicher, Bryn. "Solar on Superstores." *Frontier Group.* 2022. https://environmentamerica.org/sites/environment/files/AME-Solar-on-Superstores-1_20_22.pdf
3. "EV100 Members." *Climate Group.* https://www.theclimategroup.org/ev100-members
4. "Solar Energy Jobs." *Midwest Renewable Energy Association.* https://www.solarenergy.jobs/
5. "Careers in System Design." *Solar Energy International.* https://www.solarenergy.org/careers-in-solar/

Chapter 20- Preventative Infrastructure

1. "Infrastructure definition." *Dictionary.com.* https://www.lexico.com/definition/infrastructure
2. "Urban Heat Islands." *Climate Central.* July 14, 2021. https://medialibrary.climatecentral.org/resources/urban-heat-islands
3. "Heat Islands and Equity." *Environmental Protection Agency.* April 18, 2022. https://www.epa.gov/heatislands/heat-islands-and-equity
4. "About Urban Heat Islands." *National Integrated Heat Health Information System.* https://nihhis.cpo.noaa.gov/Urban-Heat-Islands/Understand-Urban-Heat-Islands
5. Popovich, Nadia, and Flavelle, Christopher. Summer in the City is Hot, but Some Neighborhoods Suffer More." *New York Times.* Aug. 9, 2018. https://www.nytimes.com/interactive/2019/08/09/climate/city-heat-islands.html
6. Plummer, Brad, and Popovich, Nadia. "How Decades of Racist Housing Policy Left Neighborhoods Sweltering." *New York Times.* Aug. 24, 2020. https://www.nytimes.com/interactive/2020/08/24/climate/racism-redlining-

cities-global-warming.html

7. "Coastal Risk Screening Tool." *Climate Central.* https://coastal.climatecentral.org/

8. "Study: U.S. affordable housing exposed to coastal flood risk projected to triple by 2050." *Climate Central.* May 18, 2021. https://www.climatecentral.org/press-release-affordable-housing

9. Wing, Oliver, *et al.* "Inequitable patterns of U.S. flood risk in the Anthropocene. *Nature Climate Change.* Feb. 1. 2022. https://www.nature.com/articles/s41558-021-01265-6?utm_medium=affiliate&utm_source=commission_junction&utm_content=en_textlink&utm_campaign=3_nsn6445_deeplink_PID100016836&CJEVENT=70ea29ba82ea11ec838e02630a1c0e12

10. "Coastal Risk Screening Tool." *Climate Central.* https://coastal.climatecentral.org/

11. "Heat Islands and Equity." *Environmental Protection Agency.* April 18, 2022. https://www.epa.gov/heatislands/heat-islands-and-equity

12. "About Urban Heat Islands." *National Integrated Heat Health Information System.* https://nihhis.cpo.noaa.gov/Urban-Heat-Islands/Understand-Urban-Heat-Islands

13. Dyer, Mary H. "Mangrove Tree Roots—Mangrove Information and Mangrove Types." *Gardening Know-How.* https://www.gardeningknowhow.com/ornamental/trees/mangrove/mangrove-information.htm

Chapter 21- Social Justice & Climate Justice resources

(See Chapter for compilation of Resources)

Image Credits

Introduction by Author
"Holding Up the World."
Photo image. ©Patrice Ward, 2013.

Throughout Each Chapter Section
"Air/Heaven." Photo image. ©Simon, Pixabay, June 2016

"Water." Photo image. ©Claudia, Pixabay, September 2015.

"Earth/Soil-sprouts." ©Jame, Pixabay, February, 2021.

"Fire." Photo image. ©Suhas Rawoool, April 2017.

Chapter 1
"Bell Tower."
Photo image. ©Uroborus, 2017. Pixabay.

"September 20 Global Climate Strike in Monterey, CA."
Photo image. ©Ken McLeod, 2019.

"September 20 Global Climate Strike in Monterey, CA."
Photo image ©LaVerne McLeod, 2019.

"Call at Home Stay." ©Alexa Photos, 2020. Pixabay.

Chapter 2
"Two Girls Awaiting Help-Hurricane Katrina."
Photo image. © Ted Jackson/The Times-Picayune/Landov, 2005.

Chapter 3
"Communication."
Photo Image. ©Gerd Altmann, 2016. Pixabay

Chapter 4
"Justice."
©Open Clipart-Vectors from Pixabay.

"Dr. Keeling Measuring CO_2
Photo Courtesy of Scripps Institution of Oceanography,
UC San Diego.

"Keeling Curve Graph."
Courtesy UC San Diego, Scripps Institution of
Oceanography.

Chapter 5
"Electric Charge." Photo image.
©Paul Brennan, Pixabay, July 2016.

Chapter 6
"Jail."
Image ©Kerbstone from Pixabay.

"Plastic Waste."
Photo image. ©Tokakov, Pixabay, February 2020.

Chapter 7 (no images)

Chapter 8
"Forest Tree Trunks/Deforestation."
Photo image ©Katharina N. Pixabay, August 2020.

Chapter 9
"Wind Turbines." Photo image ©Makunin, Pixabay,
June 2013

Chapter 10
"Washed out neighborhood road in Big Sur." Image
©LaVerne McLeod 2017.
"Mud Creek slide." Aerial image. ©KCBX News, 2017.

Chapter 11
"Coral Bleaching."
Photo image courtesy NOAA photo library.
©Burdick. Mariana Islands, Guam. December 2018.

"Skin Bleaching." Photo Image ©Pinetrest.com

Chapter 12
"Comparison-Larsen Ice Sheet Breakage along the Antarctic Peninsula."

Courtesy ©NASA Earth Observatory. January 16, 2022

"Comparison- Larsen Ice Sheet Breakage along the Antarctic Peninsula."

Courtesy ©NASA Earth Observatory. January 26, 2022.

Chapter 13
"Hydroelectric Power Plant."
Photo image ©Tapani Hellman, Pixabay, January 2022.

Chapter 14
"Dryness."
Photo image. ©Friedrich Fruhling from Pixabay.

Chapter 15
"Vegetables."
Photo image. ©Norbert from Pixabay.

Chapter 16
"Aquaponics for Beginner, Teachers & Kids."
Graphic design. ©EcolifeConservation.org

Chapter 17
"Types of Biomass."
©Biomass Archives-Vibrafloor.com

Chapter 18
"Air tanker releasing fire retardant over the Big Sur Soberanes Fire." Image photo. ©Ed van Weijen, August 2016.

Chapter 19
"Sun." Photo image.
©Christine Schmidt from Pixabay.

Chapter 20
"Transportation-City Infrastructure.
Photo image. ©QKampdchroer. Pixabay, August 2016.

Chapter 21
(no images)

Chapter 22
(no images)

www.ingramcontent.com/pod-product-compliance
Lightning Source LLC
Chambersburg PA
CBHW051712020426
42333CB00014B/958